U0187363

唐山
面向后工业转型的城市设计

刘 健 唐 燕 祝 贺 编著

清华大学出版社
北京

图书在版编目（CIP）数据

唐山 : 面向后工业转型的城市设计 / 刘健, 唐燕, 祝贺编著. — 北京 : 清华大学出版社, 2023.9
ISBN 978-7-302-64601-3

Ⅰ.①唐… Ⅱ.①刘…②唐…③祝… Ⅲ.①城市规划—建筑设计—研究—唐山 Ⅳ.①TU984.222.3

中国国家版本馆CIP数据核字(2023)第176215号

责任编辑：刘一琳
装帧设计：陈国熙
责任校对：王淑云
责任印制：杨 艳

出版发行：清华大学出版社
   网 址：http://www.tup.com.cn，http://www.wqbook.com
   地 址：北京清华大学学研大厦 A 座   邮 编：100084
   社 总 机：010-83470000   邮 购：010-62786544
   投稿与读者服务：010-62776969，c-service@tup.tsinghua.edu.cn
   质量反馈：010-62772015，zhiliang@tup.tsinghua.edu.cn
印 装 者：小森印刷（北京）有限公司
经 销：全国新华书店
开 本：210mm×210mm   印 张：8.2   字 数：203 千字
版 次：2023 年 10 月第 1 版   印 次：2023 年 10 月第 1 次印刷
定 价：89.00 元

产品编号：091605-01

　　本书内容是清华大学建筑学院师生团队在"2019清华大学—麻省理工学院城市设计联合工作坊"（简称"清华—MIT城市设计联合工作坊"）中的设计研究成果。特别感谢唐山市自然资源和规划局、唐山昱邦房地产开发有限公司对联合教学和本书出版的大力支持。

序言
Foreword

自1985年夏天起，清华大学和美国麻省理工学院（MIT）共同开启的建筑规划教育领域的首次中美合作——被称为"北京城市设计工作坊"（Beijing Urban Design Studio）的联合研究生暑期教学课程，至今为止已延续三十余年。2019年的联合教学选址唐山，在双方新一代教师团队的共同努力下，采取了联合设计与互访交流紧密结合的全新教学模式，并将其纳入各自的教学体系中，标志着清华与MIT这一长期的跨文化合作项目进入新的阶段。

Since the summer of 1985, Tsinghua University and the Massachusetts Institute of Technology (MIT) jointly started the first Sino-US cooperation in the field of architecture and planning education—a joint postgraduate summer program called Beijing Urban Design Studio. It has lasted for over 30 years till now. In 2019, thanks to the efforts of the new cooperative team composed of new tutors from both sides, the joint studio conducted a design research on a site in Tangshan and practiced a new

mode of internationally joint education which is featured by the integration of joint and individual design workshop with mutual study trips, as well as the incorporation of the joint studio program into the pedagogic systems of both institutes. This marked the transition of the long-term cross-cultural cooperation project between Tsinghua and MIT into a new stage.

唐山市是中国近代工业的发祥地，在中国城市发展史上占有重要一席。发生在1976年的唐山大地震，又将其以另一种方式记入了世界城市发展史册。1990年，唐山因为在抗震救灾、重建城市、解决百万人口居住问题上做出了突出成绩，获得联合国人居中心（今联合国人居署）颁发的"人居荣誉奖"，被认为是"以科学和热情解决住房、基础设施和服务问题的杰出范例"，是我国第一个获得该项荣誉的城市。

Tangshan is the birthplace of China's modern industry and occupies an important statue in the history of China's urban

development. The serious earthquake taking place in 1976 in Tangshan made the city remarkable in the world's history of urban development in another way. In 1990, Tangshan was awarded the Habitat Scroll of Honour by the United Nations Center for Human Settlements (Habitat) for its notable achievements in city reconstruction after the earthquake of 1976. As the first Chinese city receiving this honor, it was regarded as an outstanding example of solving the issues of urban housing, infrastructure and public service with science and enthusiasm.

作为中国现代工业城市的先驱，唐山创造了城市建设的辉煌，也留下了类型多样的工业设施遗存。进入新世纪，在中国深化改革开放的进程中，唐山像其他传统工业城市一样，面临着提质转型发展的巨大挑战。经济和人口的结构性变化，资源与环境的结构性约束，都是唐山城市规划建设的重要课题。从曹妃甸新区的开发建设，到南湖地区的更新整治，再到京津冀地区的协同发展，都是唐山人民面向未来挑战的积极应对。

As pioneering Chinese city of modern industrial, Tangshan has created the brilliance of urban construction, with a variety of industrial heritages. In the new century when China has been deepening its reforms and opening up, Tangshan, like other traditional industrial cities all over the world, has been facing the great challenges of upgrading and transformative development. The restructuring of economy and population, and the constraints of resources and environment, are both important issues that deserve thoughtful studies from the viewpoint of urban planning and construction. In this regard, the development of Caofeidian New District, the renovation of Nanhu Area, and the coordinated development of Beijing-Tianjin-Hebei Region are all active response to those challenges.

本书是2019清华—MIT联合教学的成果之一。清华团队从对工业设施遗存的适应性再利用，到发扬教育传统激发新的产业活力，再到重构单位社区建设全龄友好社区，以及尊重山水要素重构当地生态网络，进行了不同视角下的设计研究探索，展现了师生们针对传统工业城市在后工业化背景下转型发展这一学科前沿议题的积极思考。

This book is one of the outputs of 2019 Tsinghua—MIT Urban Design Studio. It includes the design research products of the Tsinghua team from the four perspectives of adaptive reuse of industrial facilities and heritages, education as impetus to stimulate new industries, transformation of Danwei communities into all-age friendly communities, and reconstruction of ecological networks based on landscape elements. The exploration of design research from different perspectives shows the positive thinking of the Tsinghua team on the cutting-edge issue of the transition of traditional industrial cities in post-industrial era.

作为1985年"北京城市设计工作坊"的参与者，我对今天新一代教师团队所取得的成绩深感欣慰。祝清华—MIT联合教学课程不断取得新的进步。

As participant in the Beijing Urban Design Studio in 1985, I am deeply gratified by the achievements of the 2019 Tsinghua—MIT Urban Design Studio. Wish this educational collaboration between Tsinghua and MIT could be continuously progressive and productive in the future.

毛其智
清华大学建筑学院荣退教授

MAO Qizhi, Emeritus Professor
School of Architecture, Tsinghua University

2020年6月于清华园
June 2020, in Tsinghua Garden

目录
Contents

课程介绍
Course Introduction

2019清华大学—麻省理工学院城市设计联合工作坊
2019 Tsinghua—MIT Urban Design Studio

三十年发展
Evolution of Tsinghua—MIT Urban Design Studio

在研究生培养阶段，针对城乡规划学科学生的专业学习特点与新时代的社会新需求，不断创造平台和机会，让学生能够通过参与城市设计国际联合工作坊来提升他们的规划设计技能、国际交往能力及综合专业素养的做法，目前已经成为很多高校城市规划专业培养的重要一环，也是国内高校加大自身国际化建设的重要平台和契机。清华—MIT城市设计联合工作坊（以下简称"联合工作坊"）就是这方面的典型代表。

Nowadays, at the colleges and universities that offer graduate programs of urban planning and urban design, the internationally joint urban design studio has been broadly adopted as important component of the programs, as well as effective approach to promoting their international profile. It provides graduate students with the opportunities of improving their planning and design skills, broadening their international perspectives, increasing their trans-cultural communication capacities, and enriching their comprehensive professional accomplishments, in response to the disciplinary development of urban planning and design and the transformative development of economy and society. In this regard, the Tsinghua—MIT Urban Design Studio (referred to also as the Tsinghua—MIT Studio hereinafter) can be taken as typical example.

清华—MIT城市设计联合工作坊项目始于1985年，开创了国内高校以联合教学方式开展国际合作的先河，更是国际化合作教学的成功典范。联合工作坊每两年举办一次，两校分别有2～3位教师和10～15名研究生参与，聚焦中国城镇化发展进程中的热点问题，结合教学开展规划设计研究，至今已延续30余年，积累了丰富的、具有国际前沿学术视野的教学成果。

Initiated in 1985 right after China's opening-up and reforms, the Tsinghua—MIT Studio is a pioneering project, and a successful prototype as well, of international collaborative education of China. Taking place every two years and involving participants from both universities, i.e. 2-3 faculty members and 10～15 graduate students from each party, it focuses on the cutting-edge issues of China's urbanization development and carries out researches through design for possible solutions. In the development of over 30 years, it has achieved numerous educational outcomes that are full of academic significance in international perspective.

在2019年之前，双方的教师团队经历了两代人的交替。清华方面，从早期的朱自煊和薛恩伦等教授，到后来的毛其智、张杰和单军等教授；MIT方面，从早期的Gary Hack和John de Monchaux等教授，到后来的Dennis Frenchman和Jan Wamplor等教授。在联合工作坊开展20周年和30周年之际，两校还分别在北京规划展览馆举办研讨和展览，全面展示联合教学成果，并有两册教学成果出版。

Since its initiation, the Tsinghua—MIT Studio saw several changes of the teaching team till the year of 2019, from the Professors of ZHU Zixuan and XUE Enlun to the Professors of MAO Qizhi, ZHANG Jie and SHAN Jun at Tsinghua, and from the Professors of Gary Hack and John de Monchaux to the Professors of Dennis Frenchman and Jan Wamplor at MIT. At the moments of its 20th and 30th anniversaries, both parties organized seminars and exhibitions to celebrate its success and there were also two publications about it.

在此期间，受各种条件的制约，联合工作坊的工作方式主要是MIT学生到访中国，与清华学生共同完成为期十天左右的设计研究。对于MIT的学生而言，这是一门正式的课程，并可获得相应学分；而对清华学生而言，这只是临时召集的学术活动，既未纳入正式的教学计划，也不能获得相应的学分。这使得双方在联合工作坊的合作教学中，长期处于不对等的状态，也成为2019联合工作坊寻求创新的突破口。

However, due to the limitations of the time, the Tsinghua—MIT Studio mainly took the form of a ten-days intensive workshop, which took place normally in May when the MIT team took a studying visit to China. For MIT students, it was an officially listed course with due units; while for Tsinghua students, it was only a temporary academic activity that was not included in their study plan, with no credit being officially registered. This unequal situation remained for a long time, becoming the key breakthrough when the 2019 Tsinghua–MIT Studio was planned.

新时期变革
Reforms of 2019 Tsinghua–MIT Urban Design Studio

2015年，清华—MIT城市设计联合工作坊迎来30周年庆典。之后，双方教学团队开始谋划新一轮的人员交替，并在2018年确定人选；MIT方面的Brent Donovan Ryan副教授和副教授Lorena Bello Gomez博士，清华方面的刘健副教授和唐燕副教授成为新的教学负责人。新的教师团队经过近一年的筹划和准备，确定以传统工业城市在后工业时代的转型发展作为新时期联合工作坊的主题，对破解中国传统工业城市转型发展的困境进行规划设计思考。双方同时决定，选择唐山陡河以东的工业地区作为2019联合工作坊的设计地段，并基于场地现状特点确定了采矿与景观、单位大院与城中村、工厂建筑与工业遗产、自然与人工基础设施等四个主要议题，以期为唐山的转型发展和塑造高品质的后工业城市空间提供思路和智慧，并且综合考虑两校的学期安排，共同商定了教学大纲和教学计划（表1和表2）。

When the Tsinghua–MIT Studio celebrated its 30th anniversary in 2015, the issue of renewing the teaching team was brought on the table. After about two years of discussion, the new nomination was confirmed by both parties in early 2018, with the Associate Professors of LIU Jian and TANG Yan and the Associate Professors of Brent Donovan Ryan and Lorena Bello Gomez being the responsible of Tsinghua and MIT respectively. In the following year, the four tutors of the new teaching team worked closely with each other to achieve an agreement about the subject of the Tsinghua–MIT Studio in the new era, i.e. Post-Industrial Urbanism

表1　2019清华—MIT城市设计联合工作坊教学大纲

	清华团队	MIT 团队
课程名称	城乡规划设计系列课 2：总体城市设计	Urban Design Studio/Practicum
课号	80001224	11.307j/4.173j
学分学时	4 学分，64 学时	18 学分，84 学时
负责教师	刘健副教授，清华大学建筑学院副院长，主要研究方向为城乡规划设计、城乡规划管理、城市更新和国际比较研究 唐燕副教授，主要研究方向为城市设计、城市更新、城乡规划管理	Brent Donovan Ryan副教授，麻省理工学院城市设计与发展课题组负责人，主要研究方向为收缩城市、去工业化和气候变化背景下的城市设计政策 Lorena Bello Gomez副教授，主要研究方向为大尺度地区的基础设施建设、设计对于城镇化的作用
课程助教	祝贺，博士研究生	裘熹，博士研究生
参与学生	硕士和博士研究生 10 人	硕士研究生 9 人
教学大纲	★　**课程内容**：基于学科理论知识和规划设计技能，针对唐山陡河以东处于转型中的城市综合片区，开展大尺度的总体城市设计训练，探讨传统工业城市在后工业时代的转型发展可能路径 ★　**课程目标**：深化对城市设计理论和方法的掌握与运用，了解和掌握大尺度城市空间的认知表达和城市设计理论方法，对总体城市设计范围内具有代表性和热点关注特征的城市现象和城市环境开展详细研究，并针对特定地段进行深化设计 ★　**教学组织**：各方学生均分为4组2～3人的工作小组，同时合并形成4组4～5人的混合工作小组，共同或独立完成以下环节；①对选择的设计地段进行现场调研，完成场地踏勘；②参加集中组织的课程讲授和专题报告；③基于各组的选题开展场地分析；④基于各组的选题完成总体和详细城市设计；⑤参加中间阶段和课程结束时段分别组织的中期评图和最终评图		

Table 1: Syllabus outline of the 2019 Tsinghua–MIT Urban Design Studio

	Tsinghua Team	MIT Team
Course Name	Urban Planning and Design Studio Series II: Master Urban Design	Urban Design Studio/Practicum
Course No.	80001224	11.307j/4.173j
Credit	4 credits in 64 study hours	18 units in 84 study hours
Instructors	Liu Jian, Associate Professor of Urban Planning and Design in the Department of Urban Planning and Design and Associate Dean of the School of Architecture, with special research interests on urban and rural planning and design, urban and rural planning management, urban renewal, international comparative studies. Tang Yan, Associate Professor of Urban Planning and Design at the Department of Urban Planning and Design, with special research interests on urban design, urban renewal, urban and rural planning management.	Brent Donovan Ryan, Associate Professor of Urban Design and Public Policy and Head of the City Design and Development Group in the Department of Urban Studies and Planning, with special research interests on the aesthetics and policies of contemporary urban design with respect to current and pressing issues like deindustrialization and climate change. Lorena Bello Gomez, Associate Professor in Architecture and Urbanism at the School of Architecture and Planning, with special research interests on the territorial implications of infrastructure as catalysts for design.
Teaching Assistant	Zhu He, PhD Candidate in the Department of Urban Planning and Design.	Qiu Xi, PhD Candidate in the Department of Urban Studies and Planning
Students	10 graduate and doctorate students in the Department of Urban Planning and Design and the Department of Architecture.	9 graduate students in the Department of Urban Studies and Planning and the Department of Architecture
Overview	★ Content: Along with the training on related theoretical knowledge and design skills, the studio will work on the master urban design for the industrial zone to the east of the Dou River in Tangshan that is in urban transition, as possible approach to the urban transition of industrial cities in the post-industrial era. ★ Purpose: Focusing on the subject of the urban transition of industrial cities in the post-industrial era, the studio will highlight the building of related theories and cognitions of urban design, the application of analytical methodologies on the urban assets of the design site, and the training of urban design skills to produce an in-depth design schemes for the design site at both the macro and the micro levels. ★ Organization: For each party, all participating students will be grouped into four teams of two or three members, which will be then further grouped into four mixed teams of four or five members to complete the following sessions either collectively or individually: ① field work on the design site; ② attendance to the required lectures and seminars; ③ site analysis from the perspective of the pre-defined theme; ④ urban design schemes at both the macro and the micro levels; ⑤ presentation on both the mid-term and final reviews.	

in China, aiming at working out possible solutions to the transitional development of industrial cities in view of the transition of China's urbanization. The syllabus and the schedule of the 2019 Tsinghua–MIT Studio were also confirmed then according to the academic calendars of Tsinghua and MIT, stating that the joint studio would work on the regeneration of the industrial zone to the east of the Dou River in Tangshan as exemplary for the transition of Tangshan, a typical industrial city of China, towards a high-quality urban development in the post-industrial era (see Table 1 and Table 2). Four themes were also pre-defined in view of the actual situations of the site, that is extraction and landscape, Danwei compound and urban village, factory building and industrial heritage, and natural and artificial infrastructure.

表2　2019清华—MIT城市设计联合工作坊教学计划

教学进度	时间安排		教学地点		教学内容	
	清华	MIT	清华	MIT	清华	MIT
阶段一： 实地联合调研	2019年1月20日—24日		北京／唐山		场地调研与专题研讨 ★ 在京听取有关收缩城市、京津冀协同发展、唐山城乡发展和重点项目的讲座报告 ★ 在京参观首钢工业区更新改造的最新进展 ★ 在唐山听取唐山市自然资源和规划局的相关规划介绍 ★ 在唐山踏勘设计地段及其周边地区 ★ 在唐山参加由清华大学、麻省理工学院、唐山市自然资源和规划局、昱邦公益慈善基金会共同主办的"转型中的唐山：国际经验与地方实践"学术研讨会	
阶段二： 实地联合设计	2019年1月25日—29日		北京		调研分析与初步设计 ★ 学生混合分组，在地完成调研分析，确定专题方向并完成初步设计 ★ 在地完成第一次联合设计评图	
阶段三： 线上联合设计	2019年2月27日—4月24日	2019年2月4日—4月24日	北京	波士顿	专题研究与深化设计 ★ 以学生混合分组为单位，针对各自选定的设计专题，异地线上分别开展专题研究和深化设计，定期保持线上沟通 ★ 两校教师异地分别开展设计研究指导 ★ 通过视频会议方式异地线上完成第二次联合设计评图	
阶段四： 实地联合设计	2019年4月28日—5月8日		波士顿		完善设计与成果汇报 ★ 清华团队在地开展美国案例考察，参加有关城市再开发、设计交叉研究、城市数据应用和参与式城市更新等专题的讲座学习 ★ 以学生混合分组为单位，在地共同完善设计成果 ★ 在地完成第三次联合设计评图	
阶段五： 各自成果汇报展示	2019年5月15日—6月14日	2019年5月9日	北京	波士顿	各自完成校内教学成果汇报与展示 ★ 两校团队分别异地完成各自的校内教学成果汇报与展示	

Table 2: Program of the 2019 Tsinghua–MIT Urban Design Studio

Phasing	Schedule		Location		Contents	
	Tsinghua	MIT	Tsinghua	MIT	Tsinghua	MIT
Phase I: Joint field work	20th to 24th Jan. 2019		Beijing and Tangshan		Site investigation along with lectures and seminars ★ In Beijing: lectures on shrinking cities, coordinated development of Beijing-Tianjin-Hebei Region, urban and rural development of Tangshan, and the two key projects of Nanhu Park and Donghu Park in Tangshan; site visit to Shougang Industrial Park. ★ In Tangshan: Lectures on the city planning of Tangshan, field work on the design site and its surrounding areas, and the seminar of Tangshan in Transition: Global perspective and local experiments.	
Phase II: Joint work shop	25th to 29th Jan. 2019		Beijing		Site analysis and design proposal ★ Mixed teamwork of site analysis and first design proposal from the perspective of pre-defined themes. ★ First mid-term review on site.	
Phase III: Individual design work shop	27th Feb. to 24th Apr. 2019	4th Feb. to 24th Apr. 2019	Beijing	Boston	Theme study and in-depth design ★ Mixed teamwork of theme studies and in-depth design from the perspective of pre-defined themes, along with regular online communications. ★ Individual instructions by tutors. ★ Second mid-term review online.	
Phase IV: Joint design work shop	28th Apr. to 8th May 2019		Boston		Design completion and presentation ★ Site visit to a number of American cities by Tsinghua team. ★ Lectures for Tsinghua team on the issues of urban redevelopment, cross-cutting design research, urban data application, and participatory urban renewal. ★ Mixed teamwork to complete urban design schemes. ★ Third mid-term review on site.	
Phase V: Individual review and expose	15th May to 14th June 2019	9th May 2019	Beijing	Boston	Individual review and exhibition ★ Individual final review, and exhibition as well, of the studio outcomes by each party respectively.	

新教学探索
Experiments of the 2019 Tsinghua–MIT Urban Design Studio

2019年1月中至2019年6月中，2019清华—MIT城市设计联合工作坊按照既定的教学计划如期完成，在沿袭传统、针对双方共同商定的研究议题开展教学研究的同时，努力结合新的变化在教学上尝试做出创新探索。

From mid-January to mid-June 2019, the 2019 Tsinghua–MIT Studio progressed successfully according to the teaching program agreed by both parties. While continuing the traditions of this prestigious collaborative project of international education and conducting researches by design in line with the pre-defined themes, efforts were also made to innovate the pedagogic process which can be summarized as follows.

首先，联合工作坊被正式纳入双方学校的教学计划，在教学管理上明确了责任教师、学时安排、学分认可和经费保障，保护并提高了参与联合工作坊的双方师生的积极性，也为各自教学活动的有序开展和教学质量的稳步提升奠定了重要基础。

Firstly, the 2019 Tsinghua–MIT Studio became an officially listed course at both institutes, with clear descriptions on the nomination of instructors, the arrangement of class hours, and the due credits upon the completion of the studio, as well as necessary budget for the operation of the studio. This ensured the initiative of all the participants from both parties, laying down a solid foundation for the successful progress and the pedagogic quality of the studio.

其次，联合工作坊突破了传统上的学期时间局限，灵活利用多个假期的时间，确定包括五个阶段的教学计划，确保双方既有充足的时间实地开展联合设计研究，也有充足的时间各自相对独立地开展教学并通过线上定期沟通，再通过首尾两次互访，完成"引进来、走出去"的双向互动教学全过程（图1）。

Secondly, the 2019 Tsinghua–MIT Studio broke through the time limitation of the academic semester. By making use of the possible time in the various holidays, the teaching program was composed of five phases to make it possible for the two parties to have both collective and individual work, along with regular online communications.

Moreover, the visiting trip of the MIT team to China in the early phase of the studio and that of the Tsinghua team to the US in the later phase of the studio ensured the completion of the whole interactive pedagogic process of "inviting in and going out" (see Figure 1).

再次，联合工作坊主动开门办学，积极引入地方政府、高等院校、专业机构与开发企业、社会公众等多元力量共同参与，将理论与实践相结合，将"假"设计与"真"行动相结合，既提升了联合工作坊的社会认知度，也为联合工作坊创造了更加真实的规划设计环境，丰富了学生们的设计思考维度，增强了设计研究成果的实用性。其中，唐山市自然资源和规划局在

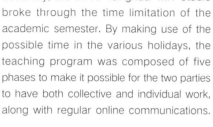

（a）

（b）

（c）

（d）

图1：2019联合工作坊的在地工作：第一阶段唐山实地调研（a）、第二阶段北京联合设计（b）、第四阶段波士顿联合评图（c）和劳伦斯社区机构座谈（d）
Figure 1: On site work of the 2019 Tsinghua–MIT Urban Design Studio: field work in Tangshan in Phase I (a), workshop in Beijing in Phase II (b), mid-term review in Boston (c), and discussion with Lawrence community organization(d)

设计地段选址、规划资料提供、场地调研安排等方面提供了大力支持，并从政府管理的角度提出了设计建议；唐山昱邦房地产开发有限公司通过昱邦公益慈善基金会给予了资金保障，并从开发商的角度提出了差异化的设计要求；来自麻省理工学院以及北京、上海、天津、唐山等地高等院校和专业机构的专家通过课堂讲座和专题研讨分享了各自的真知灼见和研究成果；唐山的社会大众是规划设计场地的长期使用者，对他们的调查和访谈构成了设计理念与构思的坚实基础；与波士顿都市区的规划部门、社区组织和企业机构等的座谈研讨，则丰富了学生们对美国城市更新的认识（图2）。

Thirdly, the 2019 Tsinghua–MIT Studio practiced the philosophy of open education by inviting representatives from governmental departments, colleges and universities, professional institutes, development companies, and local communities to join the pedagogic process in different ways, discussing together about the future development of the design site. This created a brilliant simulation of the working scenario in an actual planning and design project, helping the students to open their design perspective and to deepen their cognition on the combination between theory and practice, learning and doing. It also helped to promote the profile of the studio in the society and to increase the practicality of the studio outcomes. In details, the Municipal Bureau of Natural Resources and Planning of Tangshan provided with great helps in site selection, planning introduction, document collection, and field investigation, as well as planning and design comments from the perspective of administration. Tangshan Yubang Real Estate Development Co. Ltd provided with financial support through the donation from Yubang Foundation, as well as design consultations from the perspective of real estate development. Experts and professors from MIT and the colleges, universities, and professional institutes in Beijing, Shanghai, Tianjin, and Tangshan provided with their research outcomes and intelligent wisdoms through lectures and seminars. The public of Tangshan provided with the information about the needs and willingness of local communicates from the perspective of end users. Last but not the least, on the visiting trip to the US, the seminars with local research institutes, planning authorities, community organizations, and entrepreneurial entities also greatly enriched the knowledge of the Tsinghua team about the urban regeneration of American cities (see Figure 2).

此外，联合工作坊充分利用互联网技术，将联合设计教学从线下延伸到线上，通过网络视频会议、邮件往来沟通、社交平台互动等方式，定期开展"异地共时工作"。"在地"与"异地"相结合的共同工作模式显著提升了联合教学的广度、深度与高度，成功探索了多样化的国际联合教学的交流与共享途径（图3）。

Moreover, the 2019 Tsinghua–MIT Studio

(a)

(b)

图2："转型中的唐山：国际经验与地方实践"学术研讨会（a）和波士顿城市再开发专题讲座（b）
Figure 2: Seminar on Tangshan in Transition: Global Perspective & Local Experiments (a) and lecture on Boston Urban Redevelopment (b)

took the advanced Network technologies as regular tool to make the pedagogy be extended from reality to virtuality, especially during Phase III of Individual Design Workshop. The students of mixed groups kept regular communications with each other via virtual meetings, emails, and text chat and the two teams also organized collective workshops and mid-term review in spite of the difficulty of time difference (see Figure 3). The combination of on-site and online work helps extend the spatial and temporal scope of the studio, experimenting new and diversified possibilities of international education.

新研究议题
Themes of the 2019 Tsinghua—MIT Urban Design Studio

2019清华—MIT城市设计联合工作坊以"转型中的唐山：陡河东区"为题，针对传统工业城市的产业升级、更新改造和转型发展开展基于国际视野的规划设计思考，其中不仅涉及工业用地再开发、工业遗存再利用、城市街区再塑造等物质空间问题，还包括工人社区营造、新型产业引进、工业文化传承等社会、经济和文化议题。这既是中国众多工业城市建设发展的现实需求，也是城市规划学科针对城市更新的前沿课题。

The 2019 Tsinghua—MIT Studio was entitled as *Tangshan in Transition: Redesigning Post-Industrial Douhe East*, trying to conduct research through design with international perspectives on the transitional development of industrial cities by way of industrial upgrading and urban regeneration. It concerned not only the issues of renovating physical environment, such as redevelopment of industrial lands, adaptive reuse of industrial buildings, and reshaping of urban neighborhoods, etc., but also social issues such as re-organizing Dayuan compounds, economic issues such as introducing new industries, and cultural issues such as protecting industrial cultures. All these issues are both the actual demands of China's industrial cities and the cutting-edge problems faced by the discipline of urban planning.

唐山作为中国现代工业的诞生摇篮、震后重生的英雄城市和国家重要的重工业基地，目前正在经历艰难的产业升级和城市转型，同时也面临着京津冀协同发展的难得机遇。在此背景下，联合工作坊选择唐山陡河以东、建设中的东湖公园以西约11平方公里的城市片区作为设计研究对象，通过集聚清华和MIT两校师生的创新思维和规划设计，为唐山新时期的转型发展提供了多样化的思路。其中，清华团队分别围绕景观生态、社群组织、教育产业、基础设施四大主题，阐述了通过不同维度的更新改造激发唐山城市复兴和地区工业转型的规划设计方案思考；MIT团队则分别选择唐山热电厂、唐山钢铁厂、地段北部已拆迁城中村三个不足1平方公里的具体地块，开展了基于地段认知、功能策划、形态塑造、工业遗产保护和利用等的详细设计方案。

（a）

（b）

图3：联合工作坊第二阶段在地评图（a）和第三阶段异地线上评图（b）
Figure 3: First mid-term review on site in Beijing in Phase II (a) and second mid-term review online in Phase III (b)

Tangshan is the cradle of China's modern industry and has been one of China's key production bases of heavy industry. It became worldwide well known in 1976 due to a seriously destructive earthquake, which took away over 240,000 lives and destroyed 90% of the city's buildings, as well as a quick recovery afterwards with great supports from all over China. Currently, it is facing with both the challenges of industrial upgrading and urban transition and the great opportunities thanks to the national policy of coordinated development of BeiJing-TianJin-Hebei Region. Under this circumstance, the 2019 Tsinghua—MIT Studio focused on the site of Douhe East, an industrial zone of about 11 square kilometers to the east of the Dou River and to the west of the Donghu Park, aiming at searching for diversified approaches to post-industrial development for the transition of Tangshan, by integrating the innovative thinking and the creative design of the Tsinghua and MIT teams. In line with the four themes pre-defined in the syllabus, the four groups of the Tsinghua team put forward their master design proposals from the perspectives of landscape-ecology, community organization, educational industries, and infrastructure-oriented development respectively, while the four groups of MIT team put forward their detailed design proposals for the specific sites of Tangshan Heating Plant, Tangshan Steel-Iron Factory, and certain urban villages respectively.

新教学成果
Outcomes of the 2019 Tsinghua—MIT Urban Design Studio

一个组织良好的国际教学工作坊的成果产出往往是多层次、多维度、多类型的。2019联合工作坊传承了过去30余年清华—MIT城市设计联合工作坊的优秀传统，同时根据时代变化大胆突破创新，以取得更切合时代需求、更丰富多元的教学成果。其中既有以图纸和模型为载体的规划设计方案、设计成果展览等传统形式的教学成果，还有专题学术研讨、教学改革试点、学术共同体构建和媒体平台宣传等创新形式的教学成果。从教学的角度看，联合工作坊通过线上线下相结合的灵活教学安排，实现了国内外学分与学时的有效对接和实操性安排；通过与国内外高校、地方政府、企业单位、社会机构、城市公众等多方面的深入交流和广泛合作，改变了传统上局限于校园的"闭门教学模式"；通过双方的密切合作，在不同文化之间、不同利益主体之间、理论学习与设计实践之间搭建起联系的桥梁，促进了教学、研究与实践之间的有效衔接和良性互动，推动了跨文化、跨领域、跨角色的交流与思考，显著提升了学生的国际视野、专业知识和综合技能，是推动符合时代发展特征的高校教育国际合作教学改革的一次有益尝试。

It is quite usual that a well-organized internationally joint studio can see a rich productivity in various forms, which is verified again by the 2019 Tsinghua—MIT Studio. While continuing the great traditions of the Tsinghua—MIT Urban Design Studio accumulated over 30 years, it took the initiative to make innovations in response to the changes of the time and to produce diversified outcomes that can better meet the needs of the time, including both traditional ones, such as models and drawings of design proposal which were finally exposed on a public exhibition, and innovative ones, such as the organization of public seminars, the experiment of pedagogic reforms, the building of academic communities, and the promotion on social medias. From pedagogic perspective in particular, the realistic teaching program ensured an efficient cooperation between the two parties, the combination of on-site and online education ensured the feasibility of the teaching program, and the involvement of various members from local governmental departments, colleges and universities, institutes and companies, social organizations, and the public facilitated the openness of higher and professional education and promoted the connection between theoretical studies and practical designs. The close collaboration between the two partners of Tsinghua and MIT built up a linkage among different cultures and multiple stakeholders, as well as among education, research and practice, which greatly helped the students to broaden their international perspectives, to increase their professional knowledge, and to strengthen their comprehensive capacities through trans-cultural, trans-disciplinary, and trans-role thinking and communications. In this regard, the 2019 Tsinghua—MIT can be taken as a referential experiment of internationally collaborative higher education reform echoing to the development of the time.

刘健，唐燕
LIU Jian, TANG Yan

2020年6月于清华园
June 2020, in Tsinghua Garden

附：2019清华—MIT城市设计联合工作坊学生名单（按姓氏拼音排序）
Students participating to the 2019 Tsinghua-MIT Urban Design Studio

清华大学：   Tsinghua

郝恩琦   HAO Enqi

侯予谦   HOU Yuqian

刘 艺   LIU Yi

孟祥懿   MENG Xiangyi

邵旭涛   SHAO Xutao

许 可   XU Ke

余旺仔   YU Wangzai

张东宇   ZHANG Dongyu

张 璐   ZHANG Lu

朱仕达   ZHU Shida

MIT：   MIT

陈飞樾   CHEN Feiyue

Joude Enaam Mounir El-Mabsout   Joude Enaam Mounir El-Mabsout

Stephen Migliore Erdman   Stephen Migliore Erdman

范庄媛   FAN Zhuangyuan

Melissa Gutierrez Soto   Melissa Gutierrez Soto

Dylan Christopher Halpern   Dylan Christopher Halpern

Charlotte Xin Yun Ong   Charlotte Xin Yun Ong

Tanaya Srinivasakrishnan   Tanaya Srinivasakrishnan

王皓宇   WANG Haoyu

唐山市介绍
Brief Introduction to Tangshan City

唐山作为曾经的中国近代工业摇篮、震后重生的英雄城市和国家重工业基地，在新时期正经历着艰难的转型升级，产业结构、土地资源、人才吸引等问题成为当前唐山可持续发展的拦路虎。同时京津冀协同发展不断深入、工业文化资源持续挖掘等积极因素也正在不断显现，唐山未来发展仍大有可为。

Tangshan, once the cradle of modern Chinese industry, a heroic city reborn after the earthquake and a national heavy industry base, is undergoing a difficult transformation and upgrading in the new period. Problems such as industrial structure, land resources and talent attraction have become obstacles to the sustainable development of Tangshan. At the same time, positive factors such as the deepening of the coordinated development of Beijing, Tianjin and Hebei and the continuous mining of industrial and cultural resources are also emerging. Tangshan still has much to do in the future.

近代，唐山是中国现代工业的摇篮，这里诞生了中国第一个机械化煤矿、第一条标准轨距铁路、第一辆蒸汽机车和第一桶机制水泥。金山铁路建于1881年，是中国第一条独立修建的铁路。唐山丰富的工业遗产当前成为了最宝贵的城市财富，当地政府正在积极利用这些老工业空间实现重工业向文化产业的转型。

In modern times, Tangshan was the cradle of China's modern industry, where China's first mechanized coal mine, the first standard gauge railway, the first steam locomotive and the first bucket of machine-made cement were born. Jinshan railway, built in 1881, is China's first independently built railway. Tangshan's rich industrial heritage has now become the most precious urban wealth, and the local government is actively using these old industrial spaces to realize the transformation from heavy industry to cultural industry.

新中国成立后，唐山作为京津唐工业基地的重要一极，有钢铁、机械、化工、电子、纺织等工业，它曾经被规划为我国第二大综合性工业基地；也是北方最大的综合性工业基地。京津唐地区位于我国东部沿海地带的北部，扼我国北方地区海上门户，与日本和韩国联系便捷，拥有天津港、京唐港、曹妃甸港、秦皇岛港等港口，区位优势明显。唐山在国内外因为大地震被广为人知。1976年，唐山发生大地震，

设计场地中陡河航拍
Aerial photography of the DouheRiver in the design site
来源：720云VR全景及原宇宙创作工具平台

造成24万多人死亡。地震后，唐山在中央政府的大力支持下迅速恢复和重建。

After the founding of the People's Republic of China, Tangshan, as an important pole of the Beijing-Tianjin-Tangshan(BTT) industrial base has industries such as steel, machinery, chemical industry, electronics and textile. It was once planned as the second largest comprehensive industrial base in China. It is also the largest comprehensive industrial base in the north. The BTT region is located in the northern part of China's eastern coastal zone. It is the gateway to the sea in the northern part of China. It has convenient links with Japan and South Korea.It has ports such as Tianjin Port, Beijing-Tangshan Port, Caofeidian Port and Qinhuangdao Port, with obvious location advantages. Tangshan is widely known at home and abroad because of the great earthquake. In 1976, a major earthquake occurred in Tangshan, killing more than 240,000 people. After the earthquake, Tangshan was quickly restored and rebuilt with the strong support of the central government.

近年来，随着中国经济发展方式的转变和社会经济的发展，京津冀城市群的概念逐渐取代了京津唐工业基地的提法。因此，融入京津冀协同发展对唐山来说是一次新的机遇。2004年开始唐山经济同全国全省一样，进入高速增长的阶段，GDP超过石家庄，之后一直保持全省首位。2008年金融危机背景下，GDP增速开始下降。2012年GDP增速开始低于石家庄，GDP总量差距减小。重化工业环境严峻。宏观经济下行探底，钢铁消费市场收缩，景气度偏冷；钢铁冶金、建材等投资下降，盈利下降。环境问题突出，唐山PM2.5污染程度居于京津冀城市的中前列。

In recent years, with the transformation of China's economic development mode and the development of social economy, the concept of Beijing-Tianjin-Hebei(BTH) urban agglomeration has gradually replaced the concept of Beijing-Tianjin-Tangshan industrial base. Therefore, it is a new opportunity for Tangshan to integrate into the coordinated development of Beijing, Tianjin and Hebei. Since 2004, Tangshan's economy, like that of the whole province, has entered a stage of rapid growth. GDP surpasses Shijiazhuang and has remained the first in the province ever since. Under the background of the 2008 financial crisis, GDP growth began to decline. In 2012, the GDP growth rate began to be lower than that of Shijiazhuang, and the gap between GDP and GDP decreased. Heavy chemical industry environment is grim. The macro-economy is bottoming out, the steel consumption market is shrinking, and the prosperity is on the cold side. Investment in iron and steel metallurgy, building materials, etc. declined, while profits declined. Environmental problems are prominent, and the pollution level of Tangshan PM2.5 is among the highest in BTH region.

设计场地居住区部分航拍
Aerial photography of residential areas in the design site
来源：720云VR全景及原宇宙创作工具平台

新时期国家提出京津冀协同发展战略，核心是京津冀三地作为一个整体协同发展，要以疏解非首都核心功能、解决北京"大城市病"为基本出发点，调整优化城市布局和空间结构，构建现代化交通网络系统，扩大环境容量生态空间。京津冀协同发展是当前中国三大国家战略之一，唐山则是其中重要的节点城市。新版唐山市总体规划提出"开放融入、创新驱动、生态优先、产业多元、空间重构"五大总体发展策略，同时新版唐山市总体城市设计亦提出了"山水格局构建、文化格局构建、城市形态塑造、重点风貌分区指引"等空间发展策略。在上述背景下，积极调整城市空间结构，同时开展功能目标与空间形态相协调的综合性城市设计，已经成为促进城市深度转型的重要抓手。

In the new era, the state has put forward the coordinated development strategy of Beijing, Tianjin and Hebei. The core of the strategy is the coordinated development of Beijing, Tianjin and Hebei as a whole. It is necessary to adjust and optimize the urban layout and spatial structure, build a modern transportation network system, and expand the ecological space of environmental capacity based on the basic starting point of relieving Beijing of functions nonessential to its role as the capital and solving the "big city disease" in Beijing. The coordinated development of BTH region is one of China's three major national strategies, of which Tangshan is an important node city. The new *Tangshan master plan* puts forward five general development strategies of "opening and integration, innovation-driven, ecological priority, industrial diversification and spatial reconstruction". Under the above background, it has become an important starting point to promote the deep transformation of the city to actively adjust the urban spatial structure and to carry out comprehensive urban design with coordinated functional objectives and spatial forms.

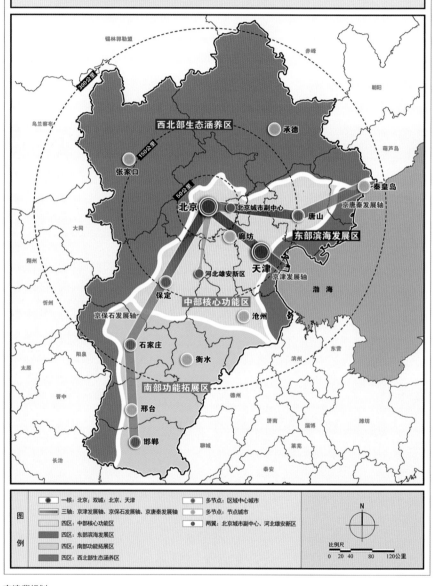

京津冀规划
BTH region planning
来源：《北京城市总体规划2016年—2035年》

场地介绍
Brief Introduction To The Site

场地区位
Site position
来源：唐山城市总体规划草案

城市设计场地位于唐山中心城区东部，属于大地震后的集中重建区。场地总面积11平方公里，规划人口12万，是一个独立的城市组团。场地现状条件复杂，新建小区毗邻大型工业企业和城中村。在最新版城市总体规划中，该区域被确定为中心城区的主要延伸发展区域，主要规划功能为住宅和综合服务。场地中的建筑陶瓷厂遗址和旧窑址等是唐山市重要的工业文化遗址，未来唐山陶瓷文化博览园将在此建成。陡河作为场地的西部边界，南北贯穿唐山市，是该地区重要的生态廊道。

The urban design site is located in the eastern part of the central city of Tangshan and belongs to the concentrated reconstruction area after the major earthquake. With a total area of 11 square kilometers and a planned population of 120,000, the site is an independent city group. The current situation of the site is complicated. The newly-built residential area is adjacent to large-scale industrial enterprises and villages in the city. In the latest version of the city master plan, this area has been identified as the main extension development area of the central city, with the main planning functions of residential and comprehensive services. The ruins of the ceramics factory and the old kiln in the site are important industrial relics of Tangshan, where the Tangshan Ceramic Culture Expo Park will be built in the future. Douhe River, as the western boundary of the site, runs through Tangshan City from north to south and is an important ecological corridor in the region.

基地北部的上村是尚未被拆除的城中村。场地中的唐山钢铁厂、唐山发电厂正在等待搬迁。陡河畔的瓷源湖公园是场地范围内最大的公共空间，也是唐山市公园绿地系统的重要节点，占地6.4公顷。在唐山工业文化格局总体规划方案中，南起启新码头，北至唐山陶瓷文化博览园的文化探访线路将串联起上述所有节点空间。此外，位于场地东侧的东湖地区将利用原有煤矿塌陷区，形成唐山市最大的生态郊野公园。

Shangcun village in the north of the site is a village in the city that has not yet been demolished. Tangshan Iron and Steel Plant and Tangshan Power Plant are removing to other locations. The former plants as important industrial and cultural sites in Tangshan city will be kept and renewed. Porcelain Source Lake Park on Douhe River is the largest public space in the site, and is also an important node of Tangshan's park green space system, covering an area of 6.4 hectares. In the overall planning of Tangshan's industrial and cultural pattern, all the above-mentioned node spaces will be connected in series by the cultural visit route from the south to Qixin Dock and the north to Tangshan Ceramic Culture Expo Park. In addition, the East Lake area located on the east side of the site will use the original coal mine subsidence area to form Tangshan's largest ecological country park.

祝贺
ZHU He

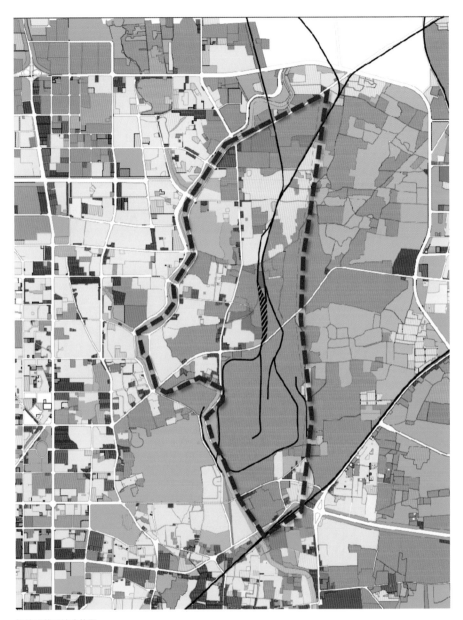

场地现状用地功能图
Current land use of site
来源：唐山城市总体规划草案

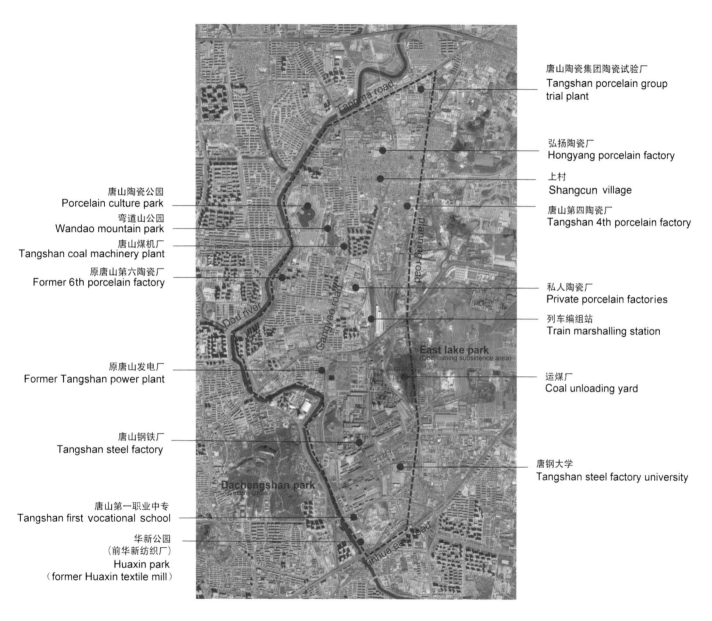

唐山陶瓷集团陶瓷试验厂
Tangshan porcelain group
trial plant

弘扬陶瓷厂
Hongyang porcelain factory

上村
Shangcun village

唐山第四陶瓷厂
Tangshan 4th porcelain factory

唐山陶瓷公园
Porcelain culture park

弯道山公园
Wandao mountain park

唐山煤机厂
Tangshan coal machinery plant

原唐山第六陶瓷厂
Former 6th porcelain factory

私人陶瓷厂
Private porcelain factories

列车编组站
Train marshalling station

Dou river

Gangyao road

planning road

Tangjia road

East lake park
(Coal mining subsidence area)

原唐山发电厂
Former Tangshan power plant

运煤厂
Coal unloading yard

唐山钢铁厂
Tangshan steel factory

Dachengshan park
(culture circle)

唐钢大学
Tangshan steel factory university

唐山第一职业中专
Tangshan first vocational school

华新公园
（前华新纺织厂）
Huaxin park
（former Huaxin textile mill）

Xinhua east road

场地要素
Elements in site

精明基础设施引导下的陡河东片区开发
Smart Infrastructure-Triggered Development

侯予谦　张东宇
HOU Yuqian　ZHANG Dongyu

背景介绍：唐山发展史与基础设施
Background introduction

唐山的建设深受大型基础设施影响，本团队从其历史背景出发，分析面临的问题，选取最重要的两项设施：工业铁路设施与绿色基础设施，提出引导"精明增长"的规划设计策略。唐山的发展与繁荣始于铁路基础设施。1879年，唐山诞生了中国第一条铁路——唐胥铁路，并由此孕育众多工矿企业，进而带动人口迁入、商业发展、城市扩张。但同时，影响唐山城市发展的基础设施不止铁路，而是随历史进程不断演变的，推动着唐山工业化设施兴建，到唐山震后重建的市政建设，再到后工业时代城市更新、生态修复的绿色基础设施。伴随基础设施演变的，是唐山城市发展模式的转变。

The construction of Tangshan is deeply influenced by large-scale infrastructures. In line with the historical background of Tangshan City, our team analyzes the problems faced, selects the two most important infrastructures – industrial railway & green infrastructure, and proposes planning and design strategies to guide the "smart development". The prosperity of Tangshan began with the railway infrastructure. In 1879, the Tang-Xu (Tangshan-Xugezhuang) Railway, as the first railway of China, triggered many industrial and mining enterprises to develop, and led to population migration, commercial development, and urban expansion. At the same time, the infrastructures that affected Tangshan's urban development were not only railways, but also the municipal infrastructure constructed after the Tangshan Earthquake as well as the green infrastructure constructed in urban renewal and ecological restoration in the post-industrial era. With the evolution of infrastructure, there is the transformation of Tangshan's urban development pattern.

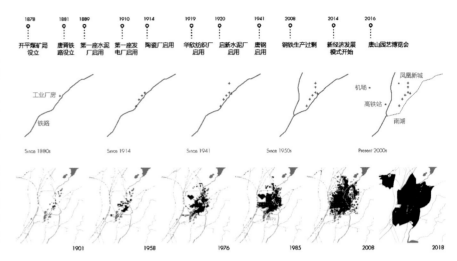

唐山城市发展时序
图源：前期工作坊合作团队
The timeline of Tangshan's urban development
Source: THU & MIT joint studio

唐山的基础设施于不同时期的定义
图源：前期工作坊合作团队
The evolution of main infrastructure in Tangshan
Source: THU & MIT joint studio

基础设施引导唐山发展：机制与问题
How does infrastructure drive Tangshan's development?

从区域关系看，以铁路为主的基础设施使唐山在区域中的产业分工与发展定位不断演变。清末民初由于唐胥铁路的建设，唐山成为当时重要的工业、矿业中心，铁路将唐山与其他以商业为主的服务型城市联系起来，如天津和秦皇岛。到20世纪50—80年代，由于铁路京哈线的建设，唐山进一步强化北京周边工业原料与产品（煤、钢铁等）运输中心的地位。近20年来，随着高速铁路的新建、重载货运铁路与深水港口的外迁，唐山中心城区重工业逐步向外围转移，区域工业地位逐步下降，同时迎来服务业发展机遇。

In regional scale, the railway-based infrastructure had made Tangshan's status in the industrial division and development orientation constantly evolving. At the end of the Qing Dynasty and the beginning of the Republic of China, Tangshan became an important industrial and mining center. The railway connected Tangshan with other service-oriented cities, such as Tianjin and Qinhuangdao. From the 1950s to the 1980s, due to the construction of the Jing-Ha (Beijing-Harbin) Railway, the status of Tangshan as the transportation center for industrial raw materials and products (coal, steel, etc.) around Beijing was further strengthened. In recent 20 years, with the construction of new high-speed railways, heavy-duty freight railways, and deep-water ports, the heavy industry in the central city of Tangshan has been gradually moved to the periphery, and Tangshan's status as the regional industrial center has gradually declined. However, in this period, the city has seen the opportunity for service industry development.

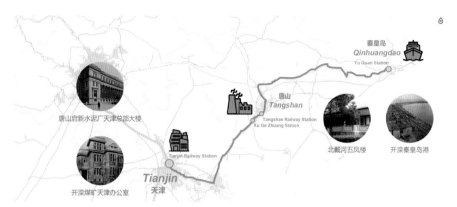

清末民初区域铁路塑造工业城市与商业城市的关系示意
Regional infrastructure and urban development pattern（Late Qing Dynasty）

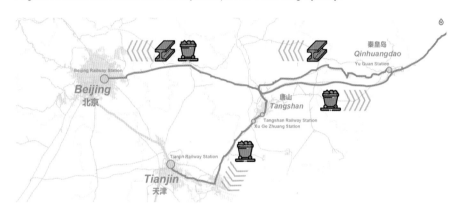

建国初期区域铁路促使唐山重工业向外扩展示意
Regional infrastructure and urban development pattern (the beginning of PR China)

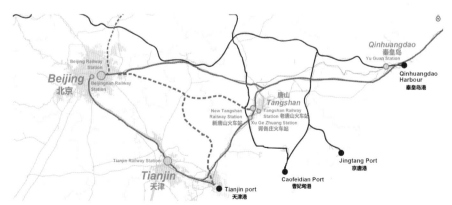

近十年新货运铁路、高铁网络促使唐山已不再是重要工业城市示意
Regional infrastructure and urban development pattern (in recent ten years)

从城市空间看，铁路基础设施使唐山形成以各大工矿企业为核心的组团式发展模式，形成了丰富的城市肌理和独特的城市形态；大规模市政基础设施引导唐山形成规避地震带的跳跃式发展模式；集中的绿色基础设施与市政设施相伴相生，共同引导城市的扩容提质与房地产主导的开发模式。这些基础设施的建设在助推唐山的快速扩张与积极转型的同时，也给唐山市区带来了城市发展与城市空间问题。

In urban scale, the railway infrastructure has enabled Tangshan to establish a clustered development pattern centered on major industrial and mining enterprises, forming unique urban texture and form; the large-scale municipal infrastructure guides a leapfrog development in Tangshan to avoid seismic belt; concentrated green infrastructure and municipal infrastructure are accompanied by each other to jointly guide the expansion of the city and the real estate led development pattern. The construction of these infrastructures has facilitated rapid expansion and active transformation of Tangshan, but it has also brought about urban development and urban space problems to Tangshan.

问题一：城市过度建设。这不仅威胁到城市安全，也使政府财政与城市运营不可持续。

Question 1: Over-construction of the city. It not only threatens urban safety, but also makes government finance and city operations unsustainable.

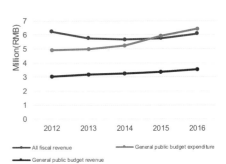

唐山市区居住用地分布
图源：唐山市城市空间发展战略研究，清华同衡规划设计研究院
Distribution of residential estate in Tangshan
Source: Research on the Development Strategy of Urban Space in Tangshan City

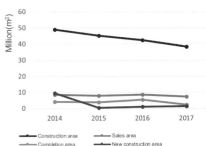

唐山市区地质灾害潜在地区分布
改绘自：唐山市城市空间发展战略研究，清华同衡规划设计研究院
Distribution of fault zones and subsidence zones in Tangshan
Source: Adapted from the Research on the Development Strategy of Urban Space in Tangshan City

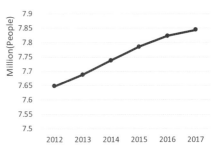

唐山市住宅交易量与财政收支变化
图源：唐山市城市空间发展战略研究，清华同衡规划设计研究院
Changes in housing transaction and fiscal revenue and expenditure in Tangshan
Source: Research on the Development Strategy of Urban Space in Tangshan City

问题二：公共服务设施过度集中。陡河东西两岸公共设施差异大，东岸成为城区的被隔离地带。

Question 2: Imparity in public service facilities. Because of imparity in the public service facilities on both banks of Douhe River, the east bank has become a spatially isolated area in Tangshan City.

问题三：绿色基础设施过度集中、品质较低。大型公共绿色基础设施集中在断层带与采煤塌陷区，与城市人口聚集区距离较远，不能成为满足市民休憩需求的公共空间，也不利于防灾。

Question 3: Over-concentration and low quality of green infrastructure. Many large-scale public green infrastructures are concentrate in fault zones and subsidence zones. They are unsafe and far from the urban area, unable to satisfy citizens' needs for leisure and rest.

问题四：基础设施导致开放空间的破碎化。大量的铁路基础设施割裂城市公共空间，城市空间品质、活力较低。

Question 4: Fragmentized open space. Tangshan's urban space is fragmented by the railway infrastructures, resulting in poor quality and low vitality of the city.

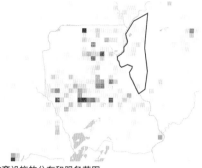

教育设施的分布和服务范围
Distribution and service area of educational facilities

医疗设施的分布和服务范围
Distribution and service area of medical facilities

生活服务设施的分布和服务范围
Distribution and service area of living service facilities

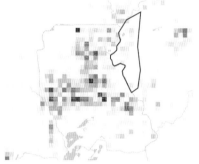

所有服务设施的分布和服务范围
Distribution and service area of all service facilities

唐山市区绿色基建分布
Distribution of current green spaces & open spaces in Tangshan City

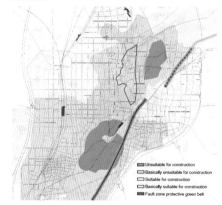

唐山市区土壤液化及塌陷区分布
图源：唐山市城市空间发展战略研究，清华同衡规划设计研究院
Distribution of soil liquefaction areas and subsidence zones in Tangshan
Source: Research on the Development Strategy of Urban Space in Tangshan City

工程思维下的基础设施建设带来不连续的城市空间
Fragmented urban spaces caused by engineering
thinking-led infrastructure construction

定位——工业组团服务中心，存量用地更新示范区
Objective: a public service center for suburban industrial clusters & a demonstrative area for the renewal of brown fields and old residential areas

唐山，这座曾因大型基础设施建设而兴起的城市，要如何再利用基础设施引导城市完成转型，实现精明发展？从市域来看，丰润、开平、古冶、曹妃甸等郊区承接唐山原有工业，但缺乏优质公共服务安置新的就业人口、引导产业升级。拥有良好工业基础和公共服务基础的唐山市区，需要成为各工业组团的生活服务与生产服务中心。

Tangshan downtown area is surrounded by industrial clusters where service facilities and high-quality residential areas are rare.

然而，在这一定位之下，唐山市区的空间扩张存在限制——市区东侧、南侧、西侧为采煤塌陷区、断裂带，北侧集中分布着干线铁路。因此需要挖掘市区存量空间（工业棕地与铁路基础设施用地）进行更新。规划范围内分布着大量存量工业棕地、铁路基础设施

Because of the rail network for commuters, Tangshan downtown area will become an important service center for the living and production in the surrounding suburban industrial clusters.

用地和附属居住用地，是重要的更新潜力地区与示范地区。

However, there are many restrictions on urban construction, such as subsidence zones and fault zones in the eastern, western, and southern parts of the city, as well as intensive railway lines in the northern part. It is difficult to expand the urban construction and there is a need to carry out renewal inside the urban area. So, infrastructure will shape the function of the city .

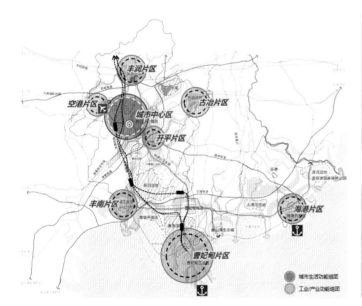

唐山范围内城市功能分布
图源：唐山市城市空间发展战略研究，清华同衡规划设计研究院
Distribution of city function in Tangshan
Source: research on the development strategy of urban space in tangshan City

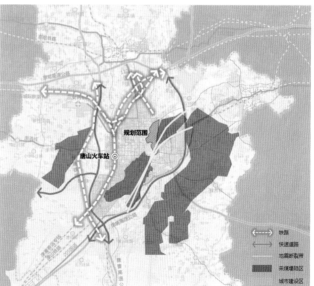

唐山市区发展限制
图源：唐山市城市空间发展战略研究，清华同衡规划设计研究院
Urban development restrictions in Tangshan
Source: research on the development strategy of urban space in Tangshan City

定位——市民文化与生态休闲的交汇处
Objective: a combination of civic cultural sites and ecological sites

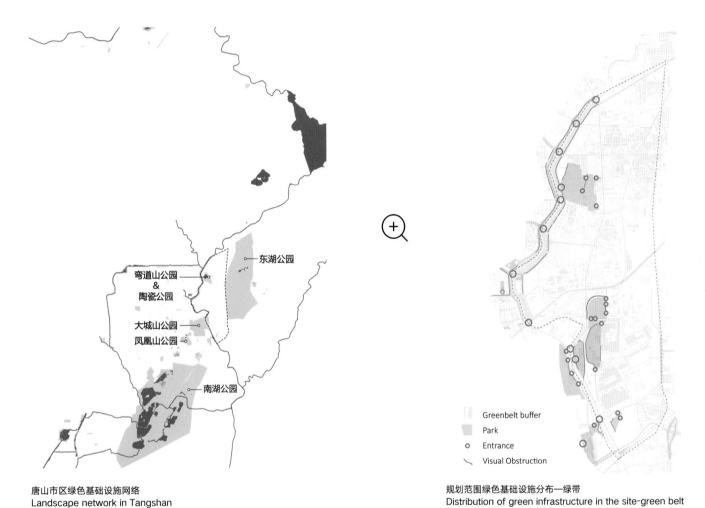

唐山市区绿色基础设施网络
Landscape network in Tangshan

规划范围绿色基础设施分布—绿带
Distribution of green infrastructure in the site-green belt

Greenbelt buffer
Park
Entrance
Visual Obstruction

规划范围位于南湖公园至东湖公园的景观廊道上，南湖公园与大城山公园聚集了较多市级文化场馆，东湖公园未来将成为重要的郊野公园，处在连接处的规划范围将成为市民文化与生态休闲融合的片区。

The site is a part of the landscape network in Tangshan. Municipal cultural venues will be gathered more in the South Lake Park & Dacheng Hill Park, and the East Lake Park will become an important country park in the future in the landscape network. The site will become an area where the civil culture and the ecological leisure are integrated to link up the landscape corridor between the South Lake Park & the East Lake Park.

定位——体现城市历史脉络的公共空间与宜居片区
Objective: a featured public site with industrial history and a cozy residential area

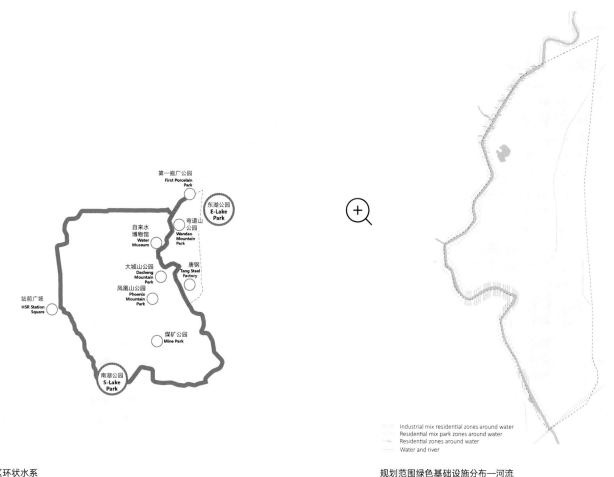

唐山市区环状水系
River ring in Tangshan

规划范围绿色基础设施分布—河流
Distribution of green infrastructure in the site: river

规划范围也位于唐山环形水系的重要节点上。环形水系串联唐山火车站、南湖公园、大城山公园、弯道山公园与凤凰新城，以大型新开发区、大型枢纽、大型公园为主。规划范围内沿河分布着大量亟待更新的单位大院，瓷厂、电厂和钢厂的厂房群，既能延续环形水系公共空间的日常性，又能打造不同于新开发区的反映城市历史、体现工业特点的特色景观与生活社区。

The site is also an important part of the river ring in Tangshan.The river ring connects many new development zones and large-scale hubs and parks. Along the river within the site, a large number of Danweis, as well as the building groups of the ceramic factory, the power plant, and the steel plant that are in need of renewal are distributed. It can create distinctive landscapes and living communities that reflect not only the city history but also the industrial characteristics, which is different from the new development zones.

035

定位——低门槛创业办公区
Objective: a low-threshold workspace for startups

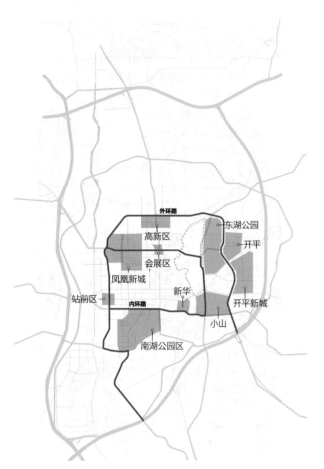

唐山市区规划快速路网
Planned expressway network in Tangshan

规划范围交通基础设施分布—道路
Distribution of transportation infrastructure in the site-road

Highway
Primary Road
Secondary Road
Service or Residential Road
Pedestrian

规划范围位于市区规划内外环快速路之间。内外环快速路沿线已有大型金融中心（凤凰新城）、高新区、会展区、商务区（站前区），然而这些开发区多用以招商引资。本地大量中小规模企业承租能力相对较弱，本规划范围可以利用类似的交通区位错位发展，成为服务本地企业与部分转型工矿企业的中小型办公区与办公居住混合区。

The site is located between the planned inner & outer ring roads. There are large-scale financial centers (Phoenix New City), high-tech zones, exhibition areas, and business district (the pre-station area) along the inner & outer ring roads, which are mainly used to attract investments. However, a large

number of small & medium-sized local companies are relatively weak in renting. Through the differentiated development of similar traffic locations, the site can become an office area or a mixed office-residential area to serve the small & medium-sized local companies.

定位——工业企业服务、科创研发、职业教育中心
Objective: a center for industrial service, technological research and development, and vocational education

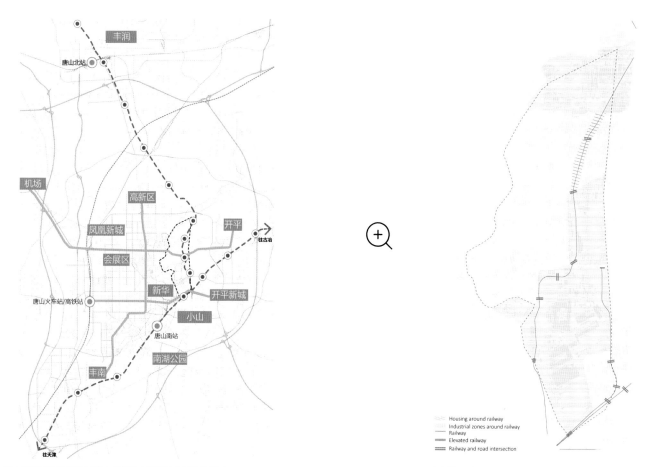

唐山市区铁路网络与规划大众运输（快速公交系统/轻轨）
Present railway network & planned mass transit system (BRT/ LRT) in Tangshan

规划范围交通基础设施分布与分类—铁路
Distribution and classification of transportation infrastructure in the site-railway

规划范围位于老京山铁路（唐胥铁路）与唐遵货运铁路的交汇点，一方面这些铁路和衍生的工业构成了地段深厚的工业积淀；另一方面这些铁路将地段与承接外迁工业的郊区组团连接起来。未来可以发挥这些铁路的客运功能，连接地段与工业组团，活化地段工业遗产与技术积淀，使地段成为与工业组团联动的生产性服务、技术研发与技职教育中心。

The site is the intersection of the old Jingshan Railway (Tang-Xu Railway) & the Tang-Zun (Tangshan-Zunhua) Freight Railway. These railways and derivative industries constitute deep industrial culture and accommodate the outward migrated industrial clusters. In the future, the passenger transport functions of these railways can be fully exerted, which can connect the site and industrial clusters and activate the industrial heritage & technological accumulation of the site, so that the site will become a center for productive service, technological research and development, and vocational education.

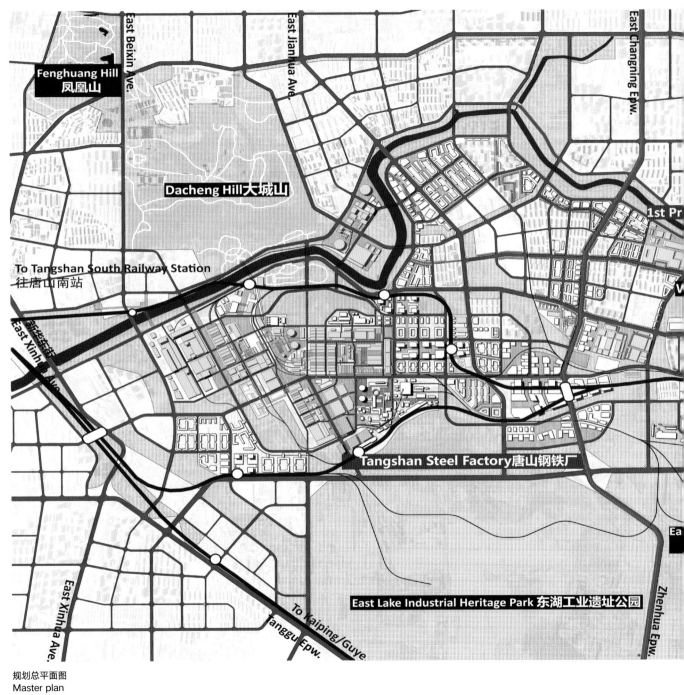

Fenghuang Hill 凤凰山

East Beixin Ave.

East Jianhua Ave.

East Changning Epw.

Dacheng Hill 大城山

1st Pr

To Tangshan South Railway Station
往唐山南站

East Xinh

Tangshan Steel Factory 唐山钢铁厂

East Xinhua Ave.

To Kaiping/Guye

Tanggu Epw.

East Lake Industrial Heritage Park 东湖工业遗址公园

Zhenhua Epw.

Ea

规划总平面图
Master plan

East Longhua Ave.

East Ronghua Ave.

Ring Epw

ories一瓷厂

Park陶瓷公园

弯道山

nd Park & Zoo

园/动物园

6th Procelain Factories六瓷厂

To Fengrun 往丰润

East Ronghua Ave.

Ring Epw

N

0   200   800

   400   1200m

总体方案策略概览
Overview on master plan strategies

铁路、市政、绿色基础设施不仅决定规划范围的区位，而且引导更新目标的确立，更是能够作为重要的空间治理手段，促进规划范围的转型。原有政府主导、大规模新建的基础设施建设模式，及其带来的大地产商主导、大拆大建的建设模式，带来了城市生态安全、空间品质与财政运行的问题。因而，需要从"基础设施网络"本身的构建与基础设施支撑土地开发两方面入手形成兼顾生态、功能合理性、空间品质、财政运营可持续的策略。基础设施网络构建重在"划定开发底线、塑造公共空间与交通骨架，并为厘清政府与开发商的开发关系提供基础"。基础设施支撑开发策略重在"街区组织、开发强度管控、空间形态引导与良性开发激励，兼顾开发利益与公共利益"。

Railway, municipal, and green infrastructures not only determine the location of the site and establish the goal of renewal, but also are important space and governance means to guide the transformation of the site. The original pattern of government-led large-scale new infrastructure construction, as well as the construction pattern of large-scale real estate and large-scale demolition and reconstruction have brought about many problems in urban ecological safety, spatial quality, and financial operation. Therefore, it is necessary to form a sustainable strategy that takes ecology, functional appropriateness, spatial quality, and financial operation into consideration from the two perspectives of establishing an "infrastructure network" and conducting land development through the support of infrastructures. The establishment of the infrastructure network focuses on "delimiting the development bottom line, shaping the public space and transportation framework, and providing the basis for defining the relationship between the government and developers". The strategy of infrastructure-supported development focuses on "block-based organization, development intensity control, space form guidance, benign development incentive, taking auount of both development interests and public interests".

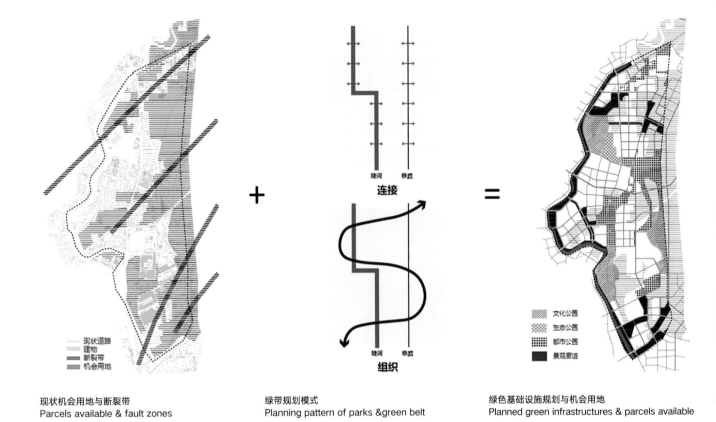

现状机会用地与断裂带
Parcels available & fault zones

绿带规划模式
Planning pattern of parks &green belt

绿色基础设施规划与机会用地
Planned green infrastructures & parcels available

绿色基础设施网络生成过程——绿带
Green infrastructure planning patterns & forms: parks & corridors

规划范围存在着由塌陷区和断裂带所带来的城市安全隐患，因而引入绿色带状基础设施作为建设区的增长边界，也将城市的景观格局进行连接与组织。

There are potential urban safety hazards brought by subsidence zones and fault zones in the site, so green infrastructure is planned as the growth boundary of the construction area, which can also be used to connect and organize the urban landscape pattern.

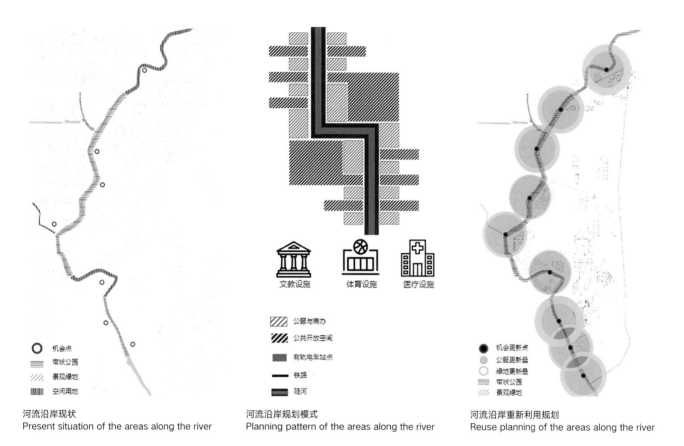

河流沿岸现状
Present situation of the areas along the river

河流沿岸规划模式
Planning pattern of the areas along the river

河流沿岸重新利用规划
Reuse planning of the areas along the river

绿色基础设施网络生成过程——河流
Green infrastructure planning patterns & forms: river

于陡河周边布置点状公共设施，并让出更多的开放空间连接东西两岸，使陡河作为西边城市和本次规划范围之间的连接与激活界面。

Public facilities are planned around the Douhe River, and more open spaces are planned to connect the east and west banks, so that the Douhe River can be used as the connection and activation interface between the western part of the city and the site.

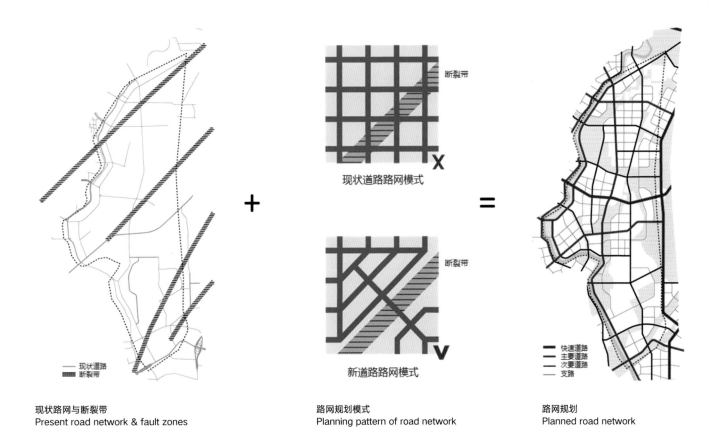

断裂带

现状道路路网模式

断裂带

新道路路网模式

—— 现状道路
▨▨▨ 断裂带

快速道路
主要道路
次要道路
支路

现状路网与断裂带
Present road network & fault zones

路网规划模式
Planning pattern of road network

路网规划
Planned road network

交通基础设施网络生成过程——路网
Transportation infrastructure planning pattern & forms: road network

在以上绿色基础设施布局基础上，将现有限制
条件（断裂带）纳入考虑，调整路网结构；并
使整体路网以小街区、密路网的形式布局。

On the basis of the green infrastructure
planning, the existing restrictions (i.e., fault
zones) will be taken into account to adjust
the road network. The overall road network
will be planned in the form of small block
and dense road network.

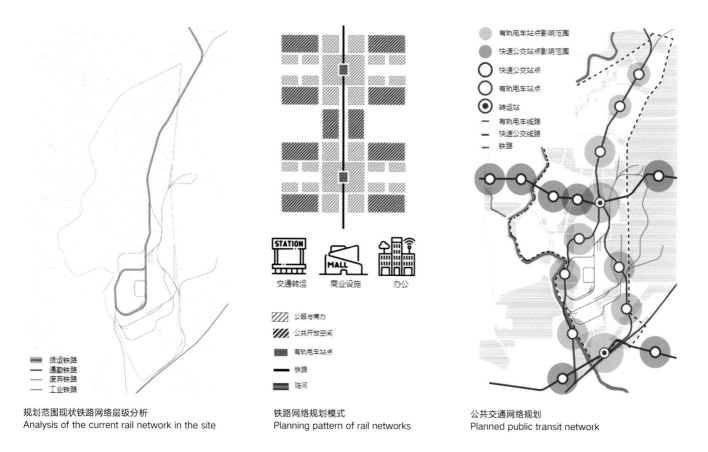

规划范围现状铁路网络层级分析
Analysis of the current rail network in the site

铁路网络规划模式
Planning pattern of rail networks

公共交通网络规划
Planned public transit network

交通基础设施网络生成过程——铁路
Transportation infrastructure planning pattern & forms: railway network

规划范围内有多条货运铁路，是唐山过去工业历史的见证，又是现今城市空间发展的障碍。未来随着以唐钢为首的工厂的迁出，这些货运铁路将有机会以有轨电车的形式得到重新利用，并结合快速公交线路促进规划范围与外围工业组团的连接，发挥通勤功能并通过车站的建设带动规划范围铁路沿线的聚集开发。

There are several freight railways in the site, which are not only the witness of Tangshan's industrial history but also the obstacle to the current development of urban space. In the future, with the relocation of factories led by the Tangshan Steel Plant, these railways will be reused in the form of tram lines. In combination with the BRT system, the tram lines can connect the industrial clusters in the periphery and the site to support the commuting and drive the concentrated development along the railway through the construction of stations.

交通基础设施如何引导城市更新、重塑城市公共生活? 一方面, 重新利用货运铁路, 将其改造为有轨电车线, 加强陡河西岸建成区、规划范围与外围工业组团的联系; 另一方面, 增设车站, 站体与公共服务设施相结合, 由此带动新组团开发与沿线大型工业厂房更新。车站周边需要进行高品质的、强度合理的聚集开发: 第一, 开发强度需要满足轨道运行的客流支撑, 按照大运量轨道交通客流支撑的经验数据, 折算出规划范围中电车车站周边新开发地块的基准容积率; 第二, 规划范围具有生态安全敏感性, 不能延续现有新开发区过高的开发强度。根据对唐山凤凰新区住宅、商办、混合用地开发强度的分析, 确定不同类型车站周边地区的大致容积率范围。在基准容积率的基础上, 通过容积率奖励制度实现容积率的增加, 鼓励新地块开发商参与公共设施完善、工业区修复更新、车站上盖开发与城市公共空间提升。

How does transportation infrastructure guide urban renewal and reshape urban public life? On the one hand, the freight railways will be reused as tram lines to strengthen the connection between the built-up areas on the west bank of Douhe River, the site, and the surrounding industrial clusters; on the other hand, stations will be built, and the new stations should be combined with public service facilities, so as to drive the development of new clusters and the renewal of large-scale industrial plants along the line. Concentrated development with a high quality and reasonable intensity should be carried out in the surrounding areas of the station: first, the development intensity should meet the needs of supporting the passenger flow of rail operation, and the benchmark plot ratio of the new development plots around the tram station in the site should be determined according to the empirical data of passenger flow supported by the mass rail transit; second, the site is sensitive to ecological safety, so the existing over-high development intensity of new development zones should not be continued in this site. On the basis of the benchmark plot ratio, the increase in the plot ratio can be realized through the plot ratio reward system, and developers of new land plots should be encouraged to participate in the improvement of public facilities, the restoration and renewal of industrial areas, the development of top cover of the station, and the improvement of urban public spaces.

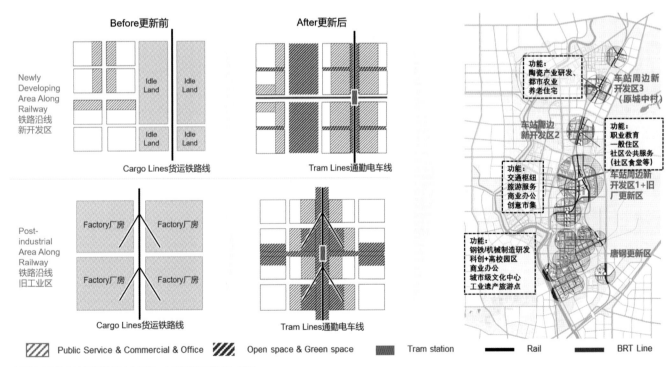

交通基础设施驱动更新的新站点开发、厂房群改造模式与规划
Development of new stations and renovation & planning of factory building groups driven by transportation infrastructure

容积率控制原则一

保证大众运输系统的客流，吸引就业与居住

地铁 – 30 000人 / 每小时断面客流
MRT – 30 000 people / hr

高架快速公交　　　　　　　有轨电车
BRT　　　　　　　　　　　TRAM

4800人
每小时断面客流
4800 people / hr

4800人
每小时断面客流
4800 people / hr

公共站点400m可达范围内人口 ≥ 8000人 (含工作与居住)

1.0
Basic FAR for
normal station areas
一般车站基准容积率

2.0
Basic FAR for hub
station areas
换乘车站基准容积率

容积率奖励一

外部配建设施/协助更新

厂房切分改造与活化
FAR 0.5+~1+

Before　　　　　　　　After

配建公园、小学等公共服务设施
FAR 0.5+~1+

Before　　　　　　　　After

车站上盖开发
FAR ≤1+

Before　　　　　　　　After

容积率控制原则二

城市中心与站点之间存在等级体系，避免过度开发

4.0/2.5
中心站商办混合区/
居住区容积率上限

2.5/2.0
外围站商办混合区/
居住区容积率上限

容积率奖励二

内部形态优化/功能混合

增加地块内公共空间

Before FAR≈2　　　　　　After FAR≈2.5

增加街道界面

Before FAR≈3　　　　　　After FAR≈3.5

功能混合

Before　　　　　　　　After

交通基础设施触发建设相关规范
Related specifications of construction triggered by grey transportation infrastructure

在绿色基础设施方面，河流与绿带周边多为公共服务设施与原有社区，间杂部分在建或待建新住宅区。对旧单位大院，可以通过绿带的渗入等方式引导地块细分、公共设施共享；对新开发地块，则缩小地块大小、提升公共空间与设施占比。

In terms of green infrastructure, public service facilities & original communities are mostly around the river and green belt, and new residential areas are under construction or to be built. For the old Danwei, the subdivision of green infrastructure and the sharing of public facilities can be guided through the infiltration of green belt; for the newly developed plot, the block size can be reduced and the proportion of public space and facilities can be increased.

落实到形态，地块开发一方面需要弥补因断裂带绿化建设而受影响的建设利益，另一方面需要保证绿带沿线的风貌，由此确定设施沿线逐级分布的开发强度管控范围。对于新开发用地，优化现有高密度住宅形态，确定滨水地块、绿带沿线地块的形态，缩小地块大小并塑造绿带渗入的公共空间，同时形成退台状的三维形态；对于旧社区更新，则整理社区空间，开辟线性步行道，并适度增加沿街底商，塑造有活力的商业与公服界面。

In the urban form, the development intensity of the block needs to make up for the construction benefits affected by the greening construction of the fault zone on the one hand, and on the other hand, it needs to ensure the landscape along the green belt, so as to determine the development intensity control scope distributed step by step along the facilities. For the new development land, optimize the existing high-density residential form, determine the form of waterfront plot and plot along the green belt, reduce the size of the plot and shape the public space infiltrated by the green belt, and form a three-dimensional space like retreating platform; for the renewal of the old community, reorganize the community space, open up a linear walkway, moderately increase the bottom business along the street, and shape a dynamic business and public service interface.

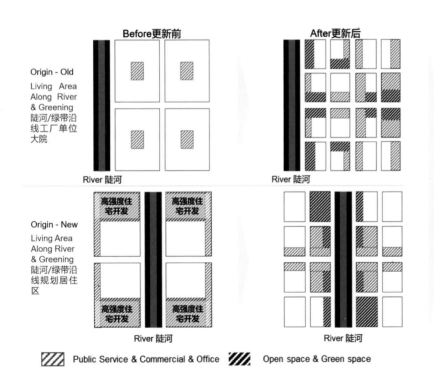

绿色基础设施驱动更新的旧社区改造、机会用地再开发模式与规划
Renovation of old communities, avalible land reconstruction mode & planning driven by green infrastructure

容积率控制原则一
补偿绿带建设损失的开发权

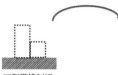

断裂带控制绿
带拆除开发量

景观带沿线
开发

49公顷
现状存在建设的规划绿带区

70公顷
绿带沿线新开发土地

70%

相对周围开发容积率可上浮幅度
(唐山多层住宅社区容积率约为1~1.2)

新开发用地规划调整

Before

After
TYPE 01

大地块 (420m*280m) 高容积率
功能单一缺少公共空间

小地块 (130m*130m) 中容积率
公建商办为主, 配部分混合社区

TYPE 02

小地块 (130m*200m) 中容积率
公建、商办、居住混合

TYPE 03

小地块 (130m*200m) 中容积率
居住、办公社区为主

容积率控制原则二
保证充足公共空间与良好景观

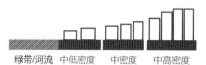

绿带/河流　中低密度　中密度　中高密度

10% ~ 30%
沿线中低密度区容积率可上浮幅度
(且沿线除少数节点外, 建筑高度不超过24m)

30% ~ 60%
沿线中密度区容积率可上浮幅度

60% ~ 90%
沿线中高密度区容积率可上浮幅度

既有社区更新

Before　　　　　　After

增加公共绿地, 减小地块

Before　　　　　　After

增加街道界面并形成合院

绿色基础设施触发建设相关规范
Related specifications of construction triggered by green infrastructure

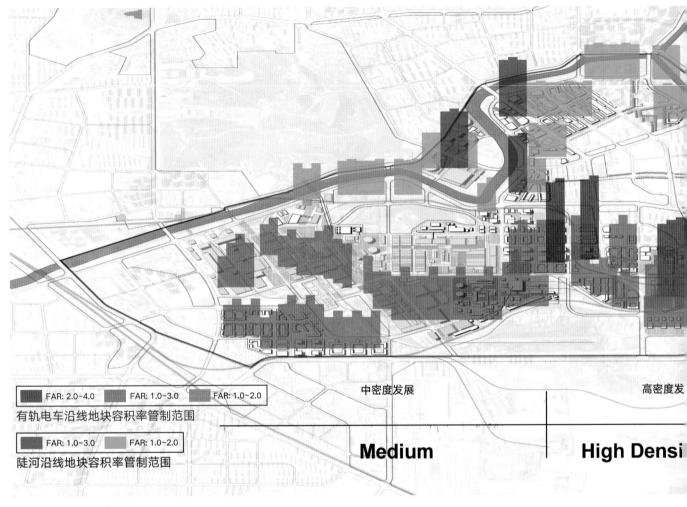

中密度发展　　高密度发

| FAR: 2.0~4.0 | FAR: 1.0~3.0 | FAR: 1.0~2.0 |

有轨电车沿线地块容积率管制范围

| FAR: 1.0~3.0 | FAR: 1.0~2.0 |

陆河沿线地块容积率管制范围

Medium　　**High Densi**

整体规划范围容积率示意
FAR in the site

通过对两类基础设施引导开发策略的分析，形成规划范围整体的开发强度与三维形态控制分区、土地使用分区规划，以及基础设施的改造重新利用与新增比例。同时，为了实现合理的开发引导，未来相关基础设施将会分阶段建设：首先是陆河以及其周边地区的整治与更新，以中部为主，并对工厂区做生态与环境修复；接着初步更新厂区周边地区；然后利用有轨电车运输重启铁路使用，并于中部与南部设置站点吸引拆迁地段的初步开发；之后开放工厂区并进行更新规划及站点设置，引导工厂更新；最后引入快速公交系统与东西城市连接，并在交汇站点设置转运站，提升站点周边的城市发展强度。

Through the analysis on the two types of infrastructure-guided development strategies, the overall development intensity and three-dimensional shape control zoning, the land use zoning planning, and the proportion of infrastructure renovation, reuse, and new increase are worked out. At the same time,

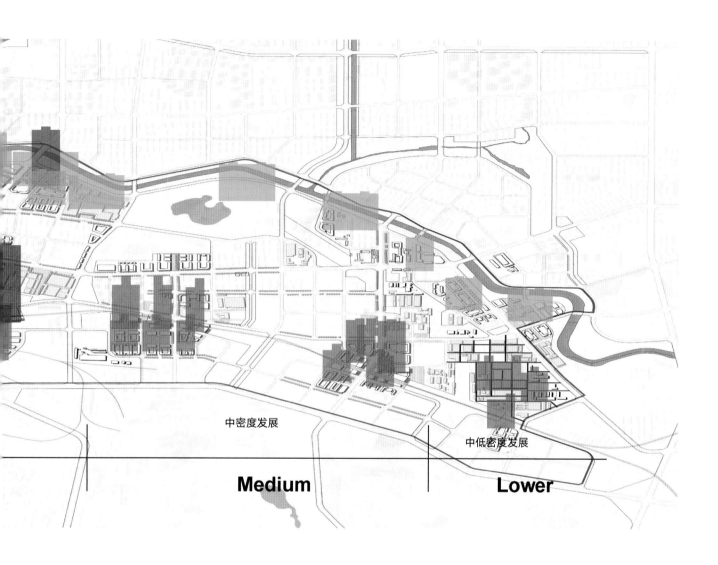

中密度发展

Medium

中低密度发展

Lower

in order to realize reasonable development guidance, relevant infrastructure will be constructed by stages in the future: the first is to carry out the renovation and renewal of the Douhe River and its surrounding areas, which is centered on the middle part and conducts ecological and environmental restoration in the factory area; the second is to carry out preliminary renewal in the surrounding areas of the factory area; the third is to reuse the railways as tram lines, and to set up stations in the central and southern parts to attract initial development of the relocation area; the fourth is to carry out renewal and build stations in the factory area to guide the factory renewal; the last is to establish the BRT system to connect the eastern and western parts of the city and set up a transfer hub at the intersection station, so as to improve the development intensity of the areas around the station.

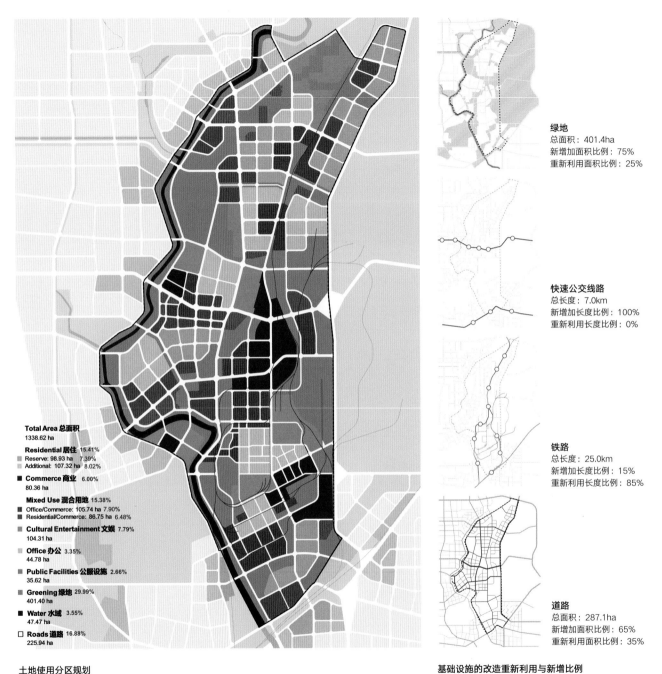

Total Area 总面积
1338.62 ha

Residential 居住 15.41%
■ Reserve: 98.93 ha 7.39%
■ Additional: 107.32 ha 8.02%

■ **Commerce 商业** 6.00%
80.36 ha

Mixed Use 混合用地 15.38%
■ Office/Commerce: 105.74 ha 7.90%
■ Residential/Commerce: 86.75 ha 6.48%

■ **Cultural Entertainment 文娱** 7.79%
104.31 ha

■ **Office 办公** 3.35%
44.78 ha

■ **Public Facilities 公服设施** 2.66%
35.62 ha

■ **Greening 绿地** 29.99%
401.40 ha

■ **Water 水域** 3.55%
47.47 ha

□ **Roads 道路** 16.88%
225.94 ha

绿地
总面积：401.4ha
新增加面积比例：75%
重新利用面积比例：25%

快速公交线路
总长度：7.0km
新增加长度比例：100%
重新利用长度比例：0%

铁路
总长度：25.0km
新增加长度比例：15%
重新利用长度比例：85%

道路
总面积：287.1ha
新增加面积比例：65%
重新利用面积比例：35%

土地使用分区规划
Land use zoning planning

基础设施的改造重新利用与新增比例
Proportion of infrastructure renovation, reuse, and
&new increase reconstruction

第一阶段
Step 1

第二阶段
Step 2

第三阶段
Step 3

第四阶段
Step 4

第五阶段
Step 5

第六阶段
Step 6

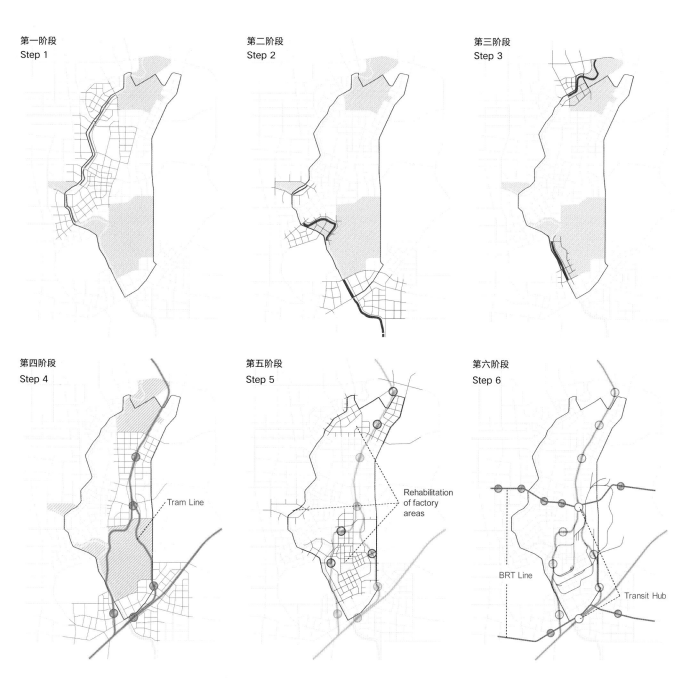

Tram Line

Rehabilitation
of factory
areas

BRT Line

Transit Hub

更新时序
Sequence of regeneration

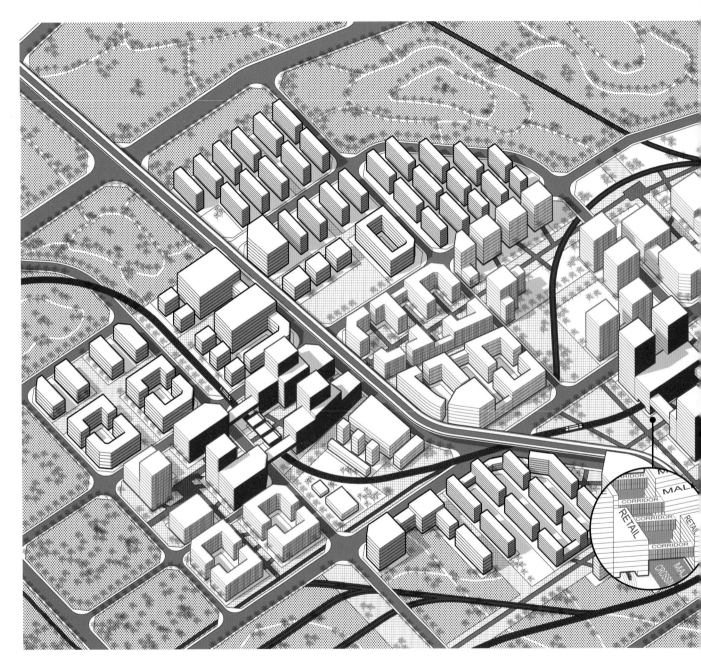

换乘站节点鸟瞰
Bird's eye view of the transfer hub area zoom-in

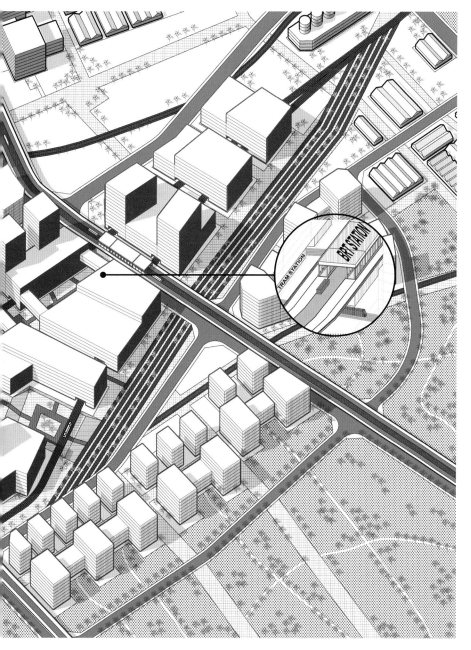

重点场地设计：换乘站节点
Hub area zoom-in

该节点位于规划范围中部东侧、唐钢厂区以北，包括由旧铁路改造的有轨电车、BRT换乘站周边地块及唐钢北侧站点周边地块。该节点除了在西侧保留部分的工厂单位住宅与厂房外，多为新建区，是典型的以交通站点吸引高强度、高公共性开发的区域。

This area is located in the east of the central part of the site and to the north of Tangshan Steel Plant area, including the land plots around the tram-BRT transfer hub and the plots around the tram station located in the north of Tangshan Steel Plant. Except some reserved factory buildings and residential buildings on the west side, most parts of the area are newly built, which can be regarded as a typical area where the high-intensity and high-commonality development are driven by transportation stations.

本团队从对公共空间的布局与品质、公共空间的立体化组织、开发强度与建筑肌理、建筑尺度与体量的思考出发，通过对该节点的详细设计进一步阐述基础设施如何合理引导新建区形成功能混合、流线顺畅、空间宜人的以公共交通为导向开发（transit-oriented development, TOD）的街区。

From the perspectives of the layout and quality of public space, three-dimensional organization of public space, development intensity and architectural texture, as well as architectural scale and volume, the team further elaborates on how infrastructure can reasonably guide the construction of new area to form TOD blocks with mixed functions, smooth transportation flow, and pleasant space through the detailed design of this zoom-area.

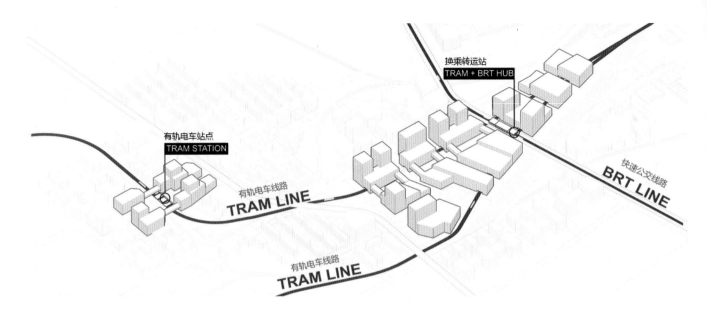

公共交通流线
Public transportation flow

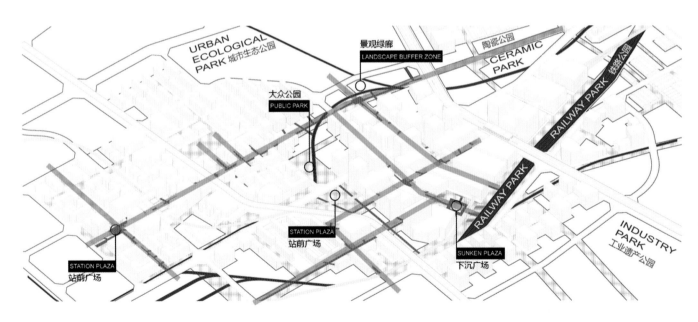

公共空间结构
Structure of public space

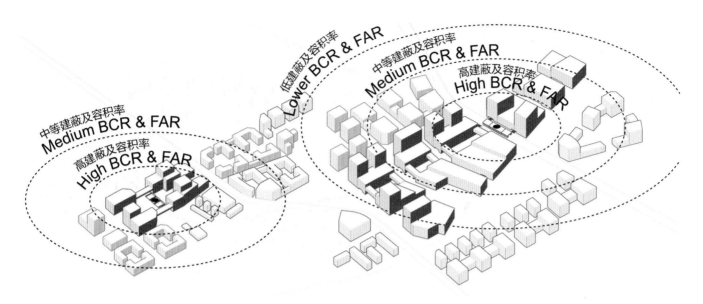

中等建蔽及容积率
Medium BCR & FAR

高建蔽及容积率
High BCR & FAR

低建蔽及容积率
Lower BCR & FAR

中等建蔽及容积率
Medium BCR & FAR

高建蔽及容积率
High BCR & FAR

圈层发展示意
Circle development

换乘站节点周边强调公共交通流线的合理组织和与土地开发的紧密结合。流线组织与公共空间形态设计上，在集结快速公交车和轻轨电车的换乘站点，以铁路上盖设计加强车站建设与商业开发的结合，并实现铁路两侧空间的联通。

In the areas surrounding the transfer hub, the reasonable organization of public transportation lines and its close combination with land development are emphasized. In terms of organizing the transportation lines and designing the public space form, in the transfer hub where BRT line and tram line are linked up, the top cover design is used to strengthen the combination of station construction and commercial development, and realize the connection of spaces on both sides of the railway.

铁路上盖将活动和功能结合，包含商场、小型商业和办公空间等业态。利用空中廊道连接站点周边非上盖建筑，并且设计主要的底层公共通道，增强步行流线的连续性。这一节点的开放空间结构从站点向外延伸，利用旧铁道车辆段设置铁道公园、下沉广场、车站广场等适合城市户外活动的公共场所。

The top cover design integrates activities and functions, including shopping malls, retail space, and office space. Air corridors are used to connect the buildings around the station. A main public passageway on the ground floor is designed to strengthen the continuity of pedestrian flow. The open space structure of this zoom-in area extends outward from the station, and old railways are used to accommodate the railway park, sunken plaza, station square, and other gathering places for outdoor public activities.

土地开发与建筑形态方面，基于TOD理论，在车站周边规划圈层发展的布局模式，开发强度从车站自内向外递减，车站的上盖实现最高强度的开发；并结合地块功能形成这一片区裙楼、塔楼、连廊、合院住宅与办公楼相辅相成的城市肌理。

In terms of land development and architectural form, based on TOD theory, the pattern of concentric development is planned around the hub. The development intensity decreases from the inside to the outside of the hub, and the top over of the hub is developed with the highest intensity. In line with the function of the site, the urban texture supplemented by the podium, tower, corridor, compound residence, and office building is formed in this area.

换乘站靠近原有铁路车辆段，在商业开发上盖建筑联通铁路东西两侧的基础上，结合建筑围合形态形成下穿铁路的下沉广场，为办公、商业建筑提供集散与活动空间；同时广场三面围合、向车辆段旧址开放，与车辆段旧址改造而成的景观绿带无缝衔接，成为车站周边最重要的公共空间。

The transfer hub is close to the original railway station. On the basis of connecting the east and the west sides of the railway through the commercialized development of the top cover, and in line with the architectural enclosure form, a sunken square is formed under the railway to provide distribution and activity spaces for office and commercial buildings. At the same time, the square is enclosed on three sides and open to the old site of the railway station. It is seamlessly connected with the green belt transformed from the old railway station through reconstruction, becoming the most important public space around the station.

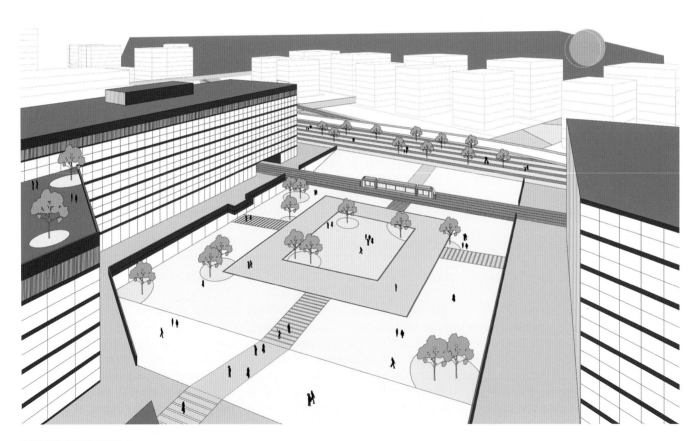

换乘站下沉广场意象效果图
The sunken square of the transfer hub

另一有轨电车车站则设置在路侧，开发强度较换乘车站稍低，无上盖开发要求，改用连廊连接铁路两侧商办建筑、形成铁路两侧的衔接。同时，这一片区的公共空间尺度较小，主要形态为垂直于铁路、衔接车站的街道与车站前小型集散广场。

The other tram station is to be built on the roadside, with a slightly lower development intensity than the transfer hub. There is no need to develop the top cover here. A corridor is used to connect commercial and office buildings on both sides of the railway.

At the same time, the scale of the public space of this area is small, mainly in the form of a street perpendicular to the railway, connecting the station, and a small square in front of the station.

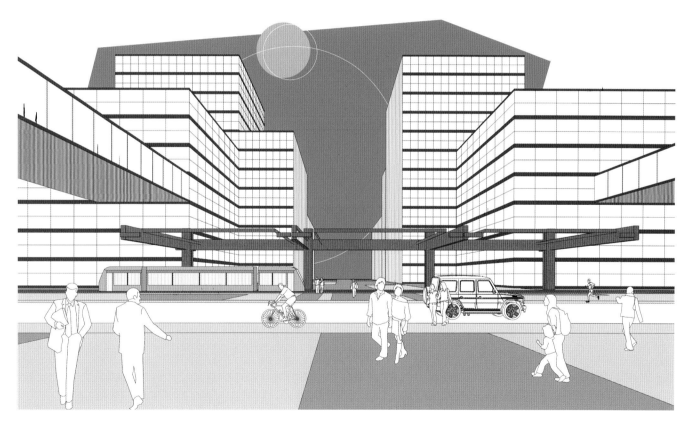

一般场站意象效果图
General station

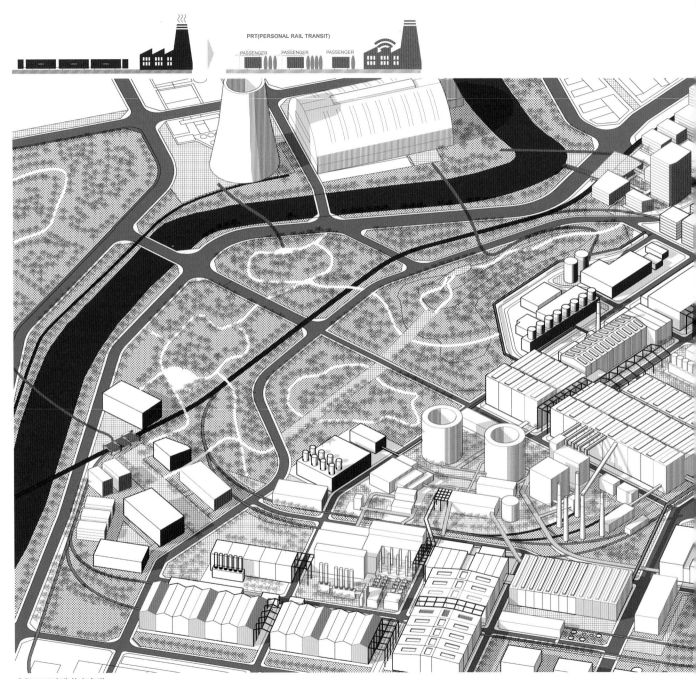

PRT(PERSONAL RAIL TRANSIT)

PASSENGER   PASSENGER   PASSENGER

唐钢工厂改造节点鸟瞰
Bird's eye view of Tangshan Steel Plant zoom-in

ATTRACTING ACTIVITIES

EXPERIMENTING TECHNOLOGIES

Mechanic-Tech   Eco-Tech

重点场地设计：唐钢工厂改造节点
Tangshan Steel Plant zoom-in

唐钢内部分布着众多铁路支线（唐遵铁路与七滦铁路唐钢专用线）、运输管道和大型钢铁工业厂房，铁路支线直接连接各大厂房，这些工业基础设施塑造了这一片区独特的城市肌理、空间氛围，是工业遗产的重要组成部分，也是城市未来的特色景观与机会空间。我们将基础设施从服务"机器"与"生产"，转变为服务"人"和"活动"，一方面利用厂房、构筑物与铁道形成服务空间、开敞空间与城市景观，另一方面改造基础设施作为技术试验与研发基地，如凭借唐山的列车制造优势产业，利用直抵建筑的铁路支线，开展PRT（个人轨道交通）实验，形成接驳电车干线的新交通模式。

There are many railway branch lines (such as Tang-Zun Railway and the Tangshan Steel Plant Special Line of Qi-Luan Railway), transit pipelines, and large-scale iron and steel industrial mills in the Tangshan Steel Plant. The railway branch lines are directly connected with major mills. These industrial infrastructures have shaped the unique urban texture and spatial ambience in this area, which is not only an important part of the industrial heritage, but also the featured urban landscape and opportunity space in the future. In the design, the function of these infrastructures is transformed from serving "machines" and "production" to serving "people" and "activities". On the one hand, mills, structures, and railways are used to create service space, open space, and urban landscape. On the other hand, the infrastructure is transformed to a base for technological experiment and research and development. For example, by virtue of train manufacturing which is the advantageous industry of Tangshan, PRT (Personal Rail Transit) experiment is developed, forming a new transportation mode that connects mills with the tram trunk line.

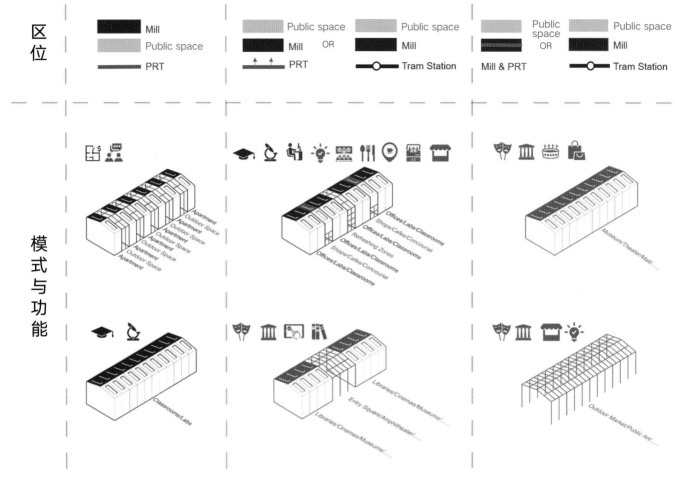

| Mill |
| Public space |
| PRT |

| Public space |
| Mill | OR |
| PRT |

| Public space |
| Mill |
| Tram Station |

| Public space |
| Public space |
| Mill |
| Mill & PRT | OR |
| Tram Station |

模式与功能

Apartment
Outdoor Space
Apartment
Outdoor Space
Apartment
Outdoor Space
Apartment
Outdoor Space
Apartment
Outdoor Space

Offices/Labs/Classrooms
Shops/Cafes/Concourse
Offices/Labs/Classrooms
Refreshing Zones
Shops/Cafes/Concourse
Offices/Labs/Classrooms

Museum/Theater/Mall/......

Classrooms/Labs

Libraries/Cinemas/Museums/......
Entry Square/Amphitheater/......
Libraries/Cinemas/Museums/......

Outdoor Market/Public Art/......

工厂场房单元改造示意
Renovation of mill units

为了将城市肌理引入大型工业厂房，将大型工业厂房根据结构逻辑分解为一个个单元，有利于道路等其他基础设施划分厂房，缩小单体改造规模利于多方参与。这些分成单元形式的工业厂房可灵活用于不同目的，基于它们与 PRT 铁路支线的关系，不同的单元组合将实现不同的功能，靠近铁路支线和 PRT 站点及轻轨车站的厂房单元的功能更多元、更面向公众开放。基于这一原则，工业厂房

被划分为不同的主题功能区，如中部原料运输与处理区改造为公共艺术中心，靠近站点的仓库改造为市场与办公组团、北侧轧钢厂群改造为研究院组团等。沿着铁路的工业厂房优先用作研究院与办公区的公共娱乐与休闲场所。同时，通过串联公共空间、周边车站及富有特色的烟囱、运输管道等装置，形成景观廊道，激发丰富的公共活动，创造城市新活力。

Aimed at integrating large-scale industrial mills with the urban texture, the mills are divided into small units according to their structural logic, which is conducive to dividing mills by roads and other infrastructures and reducing the scale of individual units for the purpose of multi-party participation. These industrial mills, in the form of divided units, can be flexibly

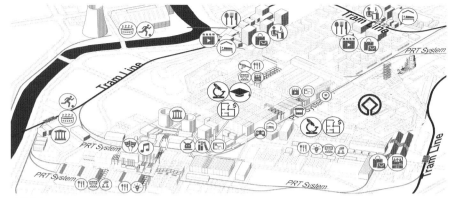

功能分布
Distribution of functions

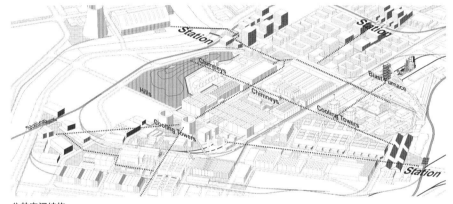

公共空间结构
Structure of public space

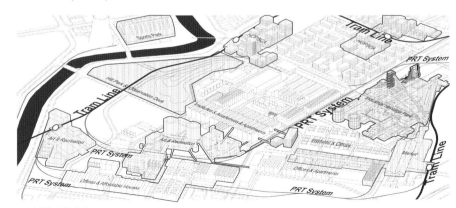

工厂建筑功能分布
Distribution of mills' functions

used for different purposes. Based on their relationship with PRT branch lines, the combination of different units can exert different functions. The mill units close to the railway branch line, PRT station, and tram station will have more diversified functions and be more open to the public. Based on this principle, the industrial mills are divided into different themed functional areas, such as a public art center transformed from the raw material transportation and processing area in the middle, a market and office area transformed from the warehouse close to the station, and a research institute group transformed from the rolling mill group on the north side. The industrial mills along the railway are transformed with priority to public entertainment and leisure places for research institutes and office areas. At the same time, the public space links up the surrounding stations, and a landscape corridor is formed by use of the distinctive chimneys, transit pipelines, and other devices, which can stimulate diversified public activities and create new vitality of the city.

厂区中央的燃料与原料运输储存区将被改造为集合艺术展陈、片区文化场馆、工业遗产活态展示与装置艺术的休闲艺术区。这里集合了大量独特的工业基础设施装置——铁路支线、筒仓、烟囱、高架运输管道等。将铁路支线改造为接驳电车车站的PRT试验系统与景观绿地；在高架管线上加建天桥，强化其视觉效果，并构建建筑之间的步行串联系统；筒仓内部改造为展陈空间，在筒仓与烟囱间增设衔接高架管廊步行系统的外挂坡道与观景平台，形成体验工业设施与装置的新场所。

The fuel and raw material transit and storage area in the center of the factory area will be transformed into a leisure and art area integrating the functions of art exhibition, regional cultural venue, display of industrial heritage, and installation art. There are a large number of unique industrial infrastructure devices -railway branch lines, silos, chimneys, elevated transit pipelines, etc. The railway branch lines are transformed into the PRT experiment system and landscape green space connecting the tram station; overpasses are added to the elevated pipelines to enhance the visual effect and construct a pedestrian connection system between buildings; the interior of the silo is transformed to a display space; an external ramp and a viewing platform are added to the silo and chimney to connect them with the elevated pipeline pedestrian system, so as to form a space for visitors to comprehensively experience and interact with industrial facilities and devices.

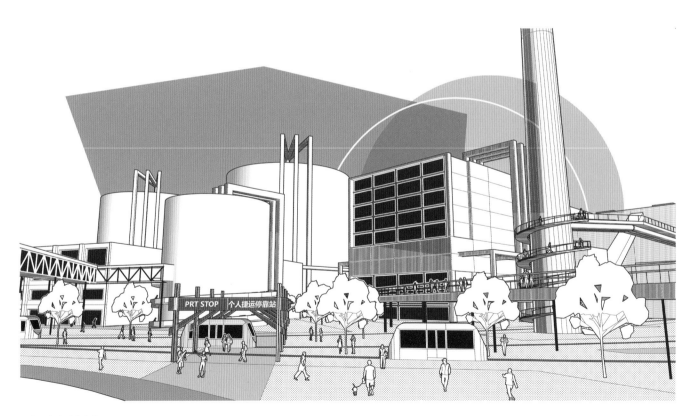

工业厂房改造意象效果图
Renovation of industrial mills

心得体会
Summary

张东宇
ZHANG Dongyu

侯予谦
HOU Yuqian

这次studio让我对城市设计有了一些不同的体验：第一次两个人完成13平方公里的总体城市设计；第一次接触专题切入式城市设计，专题切入式的城市设计需要更清晰的故事线，也由此对于研究的深度、设计的收放程度有了更高的要求。我和队友尝试构建从设施本体到空间策略、从问题提出到策略推演的逻辑关系，在这个过程中逻辑思考的能力得到了提高。这也是我第一次参加联合设计，和MIT团队共同完成前期调研，并通过联合评图进一步了解国外城市设计与规划策略的思维、工作流与表现方式，最终自己也能够在设计方案的推演与表现上将学习到的技巧与内容付诸应用。

This studio gives me some different experiences. This is the first time for me to complete an overall urban design of 13 square kilometers with my colleague and get in touch with the special themed urban design. The special themed urban design requires a clearer logistic. Therefore, it has higher requirements for the depth of research and design. In this process, my ability to think logically has been improved. This is also the first time for me to participate in a joint studio, which helps me further understand the thinking, workflow, and visualization of urban design and planning strategies of foreign countries. Finally, I am able to apply the skills learned in this joint studio in urban design & planning.

在联合设计工作坊中，可以与MIT的同学一起合作学习，是一次宝贵的机会，不仅能吸取对方对于规划设计的方法与技术，更能将这些方法应用于其中。本团队只有两人合作，过程中可以一直激发自我潜能，并很有效地将团队想法表现于设计之中，这将会是我于团队合作中最宝贵的经验与回忆。

It is a valuable opportunity. I have not only learned the methods and techniques of urban planning and design, but also applied these methods in the work of this joint studio. Our team has only two members. Through cooperation, we can always motivate our own potential in this process , and effectively present our ideas in the design. This will be my most valuable experience and memory in teamwork.

入城学院：陡河东岸新唐山
Colleges in City: A New Tangshan Image in Douhe East

郝恩琦　孟祥懿　朱仕达
HAO Enqi　MENG Xiangyi　ZHU Shida

背景介绍
Background introduction

城市发展有赖于人才，人才的培养有赖于教育机构。唐山作为中国现代工业城市，唐山交通大学（简称唐山交大）的存在对城市发展的推动有力地证明了这一论断。

As we know, urban development depends on talents and educational institutions. As for Tangshan, a modern industrial city in China, it is Tangshan Jiaotong University that drove the development of the city, which strongly proved this assertion.

唐山是中国北部一个重要的工业城市。19世纪，唐山由于丰富的矿石资源，初级的工业逐渐出现；至20世纪，工程技术水平提升，工厂数量和门类也逐渐增多，唐山工业得到了快速发展。而这一转变的支撑，是唐山交通大学的建立。

Tangshan is an important industrial city in northern China. In the 19th century, the primary industry gradually emerged due to the rich ore resources. And by the 20th century, as the engineering technology had improved, numbers and categories of factories gradually increased, since then Tangshan's industry witnessed rapid development. During this transformation, the establishment of Tangshan Jiaotong University has made great contribution.

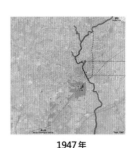

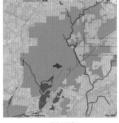

唐山"教育兴城"策略图
Strategy of "Promoting City by Education" in Tangshan

城市建成区
Urban Area

工业区域
Industrial Area

商业中心
Commercial Center

铁路
Railway

唐山城市发展用地变迁
The change of urban land use in Tangshan

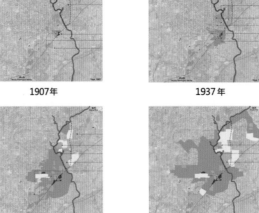

| 1907年 | 1937年 | 1947年 |
| 1971年 | 1987年 | 2017年 |

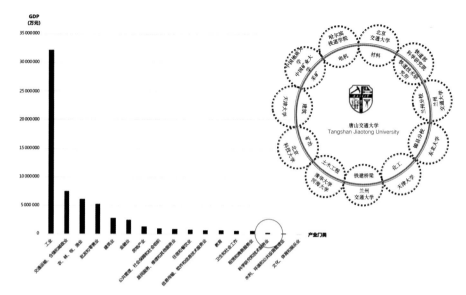

唐山市地区生产总值
Gross regional product of Tangsha

唐山交大院系调整
Faculty adjustment of Tangshan Jiaotong University

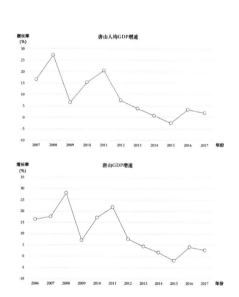

唐山GDP增速和人均GDP增速
GDP growth rate and per capita GDP growth rate in Tangshan

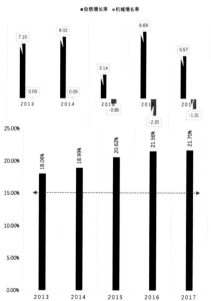

唐山人口增长率和老龄化率
Population growth rate and aging rate in Tangshan

后来由于抗战发生，唐山交大进行了颠覆性的院系调整，大量的核心技术与人才离开唐山，以至于整个唐山的信息技术、科研等高附加值行业的发展受到影响。唐山的工业经济日渐低迷，大量人才迁出，工业—教育系统遭到破坏，科研与技术教育也逐步萎缩。所以，为什么不重建唐山交通大学？它也许会成为唐山教育兴城的第一炮，让科技与研发的创新力量重回唐山，让人才与精英重回唐山。

But later, due to the Anti-Japanese War, Tangshan Jiaotong University carried out a subversive faculty adjustment, resulting in the outflow of a large number of core technologies and talents, so that the entire value-added industries were affected. At the same time, Tangshan's industrial economy was also becoming increasingly sluggish. With a large number of talents moved out, the industry-education system has been destroyed, and the scientific research and technical education have gradually shrunk as well. So, why not rebuild Tangshan Jiaotong University? Hopefully the new university may spark a boom in urban development by education and attract innovation power of research, design, and talents back to Tangshan.

意象：工业城市的文化符号

陡河东地段有40余处工业遗产

Tangshan Xinming Ceramics Factory	唐山市新明瓷厂 ①
Tangshan Hebei East Cement Co. Ltd.	唐山市冀东水泥股份有限公司 ②
Tangshan Boyu Ceramics Creative Culture Co. Ltd.	唐山博玉陶瓷文化创意有限公司 ③
Tangshan Huamei Ceramics Company	唐山华美陶瓷有限公司 ④
Tangshan Flint Ceramics Factory	唐山第一瓷厂 ④
Tangshan Xinde Boiler Group Co. Ltd.	唐山信德锅炉集团有限公司 ⑤
Tangshan Renshi Cement Co. Ltd.	唐山任氏水泥设备有限公司 ⑥
Tangshan Ceramics Group Co. Ltd.	唐山陶瓷集团有限公司 ⑦
Tangshan Xingyuan Porcelain Industry Co. Limited	唐山兴源瓷业有限公司 ⑧
Tangshan Lightbulb Factory	唐山市灯泡厂 ⑨
Tangshan Coal Mining Machinery Repair Factory	唐山市煤矿机械修理厂
Shijiazhuang Fan Tangshan Branch	石家庄市风机厂唐山制造分厂
Tangshan Xieli General Machinery Factory	唐山市协力通用机械厂
Tangshan Boiler Installation and Repair Factory	唐山市锅炉安装修理厂
唐山开尔热交换设备制造有限公司 (唐山河北冶金设备制造有限公司)	
(Tangshan Kaier Heat Exchange Manufacturing Co. Limited	
(Tangshan Hebei Metallurgical Equipment Manufacturing Co. Limited)	
Tangshan Dongyi Auto Repair Factory	唐山市鼎毅东汽车修理厂
Tangshan No. 6 Ceramics Factory	唐山第六瓷厂
Tangshan Ceramics Meishu Ceramics Factory	唐山美术瓷厂
Tangshan Leather Products Factory	唐山兴源瓷业有限公司
Tangshan Ceramics Group Mechanical Factory	唐山陶瓷集团机械厂
Tangshan Gaohua Ceramics Co. Ltd.	唐山高华国际陶瓷有限公司
Tangshan Huazincheng Ceramic Products Co. Ltd.	唐山华信成陶瓷制品有限公司
Tangshan Longying Ceramics Co. Ltd.	唐山龙英瓷业有限公司 ㉒
Tangshan Xinyv Bone China Co. Ltd.	唐山欣玉骨质瓷有限公司 ㉓
Tangshan Longhua Ceramics Oil Supply Co. Ltd.	唐山龙华陶瓷石油供应有限公司 ㉔
Tangshan Lubei Shengzao Carton Factory	唐山市第二橡胶厂综合制品厂
Tangshan Second Rubber Integrated Products Co. Ltd.	唐山市第二橡胶厂综合制品厂
唐钢新事业发展有限公司金属包装厂	
Tanggang New Development Company Metal Packaging Factory	
Tangshan Taijie Hoisting Equipment Co. Ltd.	唐山缔捷起重设备有限公司 ㉗
Tangshan Guanda Insulation Materials Co. Ltd.	唐山冠达保温材料有限公司 ㉘
Tangshan Metal Gift Box Factory	唐山市礼品盒厂
Tangshan Lubei District Non-Ferrous Casting Co. Ltd.	唐山市路北区铸造有限公司
Tangshan Non-Ferrous Metals Foundry	唐山市有色金属铸造厂

唐山市第四制陶厂 Tangshan Fourth Acid-Resistant Ceramics Factory	㉑
唐山市陈结锅炉技术有限公司 Tangshan Hongming Boiler Technology Co. Ltd.	㉒
唐山市开平区隆源新型砖厂 Tangshan Kaiping District Longyuan New Model Brick Factory	㉓
宝业集团西晓储运公司 Baoye Group Xiyao Storage Company	㉔
唐山冶金矿山机械厂 Tangshan Metallurgy Mining Machinery Co. Ltd.	㉕
唐山德博特机械有限公司 Tangshan Debote Machinery Co. Ltd.	㉖
唐山供电公司 Tangshan Power Supply Company	㉗
唐山市通用汽车修理厂 Tangshan General Auto Repair Factory	㉘
唐山市泰发汽车修理有限公司 Tangshan Taifa Auto Repair Co. Ltd.	㉙
唐钢集团艺术瓷厂 Tangshan Steel Co. Ltd.	㉚
河北大唐国际唐山热电有限责任公司 Hebei Datang International Tangshan Thermoelectric Co. Ltd.	㉛
唐山市汽车衡器厂 Tangshan Vehicle Scale Factory	㉜
唐山市礼品艺术瓷厂 Tangshan Lipin Meishu Ceramic Factory	㉝

陶瓷	CERAMICS
机械	MACHINERY
杂项制品	MISC. PRODUCTS
汽车修理	AUTO REPAIR
水泥	CEMENT
仓储	STORAGE
金属制品	METAL PRODUCTS
能源供应	ENERGY RESOURCES

陡河东岸的工业遗产优势
Industrial heritage advantages of Douhe East

环境：可持续生态品质

高质量生态环境

陡河东岸的环境和文化资源优势
Environmental and cultural advantages of Douhe East

文化联系：与交大原址的空间关联

与交大原址的文化与空间联系

实际上，唐山政府早已具有强烈的意愿重建唐山交大，近年来，唐山政府对教育的投入也非常高。所以我们将唐山交大的重建作为设计的核心策略，让其成为唐山教育之路的先行者，进而提升人口多样性、推动产业转型、重塑城市形象。面对城市退二进三的契机，唐山需要更加重视技能与科研等创新力量，以应对唐山转型乏力的核心问题。因此，一个基于新技能与科研的唐山交大复建计划由此诞生。陡河东岸拥有一些特殊优势：如工业遗产、东湖公园等。这些要素都将为唐山交大的复建提供特别的帮助：工业遗产能够为新的唐山交大提供独特的形象助益；东湖与陡河将为新校址带来阳光与绿地；而与唐交旧址的临近也为其带来涅槃的特殊意义。

In fact, local government has already had a strong desire to rebuild this foremost university. So our core strategy is to rebuild Tangshan Jiaotong University, make it become the pioneer of Tangshan's education, promote diversity of employees, assist in industrial transformation, and build up a new post-industry urban image. So what kind of university do we need? Faced with the opportunity of urban transition from secondary industry to tertiary industry and core problem of the weak transformation motivation, we put up with a redevelopment plan based on new skills training and scientific research. Given the special advantages in Douhe East, such as industrial heritage, East Lake Park, it will support the redevelopment of Tangshan Jiaotong University. Specifically, the industrial heritage could provide unique image for the new Tangshan Jiaotong University; the East Lake and Douhe River could bring sunshine and green space for the campus; and the proximity to the original campus would also bring special significance of its revitalization.

第1步 功能：寻求一个适合的新唐山交通大学体系
Step 1 Function: a suitable system of Tangshan Jiaotong University

回顾唐山交通大学（以下简称"唐山交大"）校史，20世纪90年代初因为多位毕业生在西方高校成绩极为突出，成为一所世界闻名的工程院校，教育水平得到高度认可。1916年始，凡唐山交大毕业的同学皆可免试就读康奈尔大学研究生。然而唐山交大的教育不仅限于工程领域，历史上还先后设立了技术高校、市民学校两个层级的机构，为唐山工业制造输送了大量一线人才，为工业制造的发展提供了坚实保障，开办的市民学校广泛普及拼音、识字等基础知识，帮助唐山逐步降低文盲率，提升城市的整体素质水平。唐山交大的三级教育体系为城市发展、工业发展做出了重要贡献。我们希望由此出发，并根据当今社会的需要，回应当前发展趋势，重建新唐山三级教育体系。

Looking back at the history of Tangshan Jiaotong University, in the 1910s, it became a world-famous engineering university because of the outstanding achievements of many graduates in Western universities and its education level was highly recognized. Since 1916, students graduated from Tangshan Jiaotong University can be directly admitted as graduate students to Cornell University. In addition, the education of Tangshan Jiaotong University was not limited to engineering, as it has also established educational institutions at two levels of technical schools and citizen schools in history, which provided a large number of first-line technical talents for Tangshan's industrial manufacturing, laid a solid foundation for the development of industrial manufacturing in Tangshan. The citizen schools widely popularized pinyin and literacy, gradually reduced the illiteracy rate in Tangshan and improved the overall quality of people in the city. The three-level education system of Tangshan Jiaotong University has made important contributions to urban development and industrial development, which is very far-sighted. We hope to start from this and rebuild the three-level education system according to the needs of today's society and respond to current.

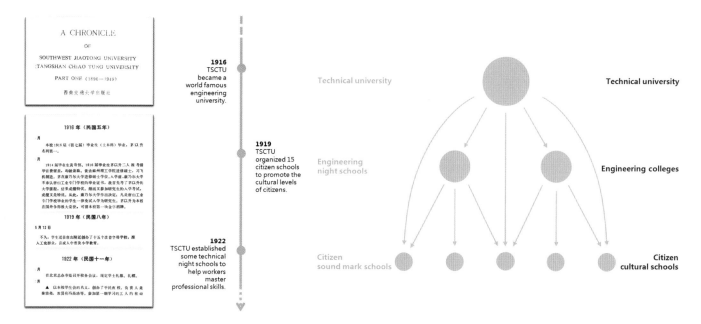

唐山交通大学大事记
Chronicle of Tangshan Jiaotong University

唐山交通大学教育体系
The education system of Tangshan Jiaotong University

Technical university	Engineering colleges	Citizen cultural schools

Ambition	High-tech talents	New professionals and re-educators	Quailfied citizens

| Spatial | Perfect technical facilities | Convenient career opportunities | Rich teaching content |
| Demand | Tranquil academic space | Adequate vocational training | Vigorous learning environment |

唐山交通大学新三级教育体系
The new three-level education system of the Tangshan Jiaotong University

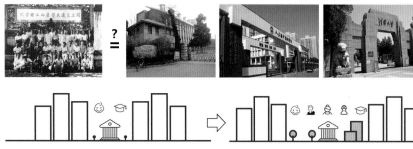

Gated School – The Knowledge Palace   Knowledge Palace with Diversified People and Vitality!

唐山交通大学校园空间模式
Campus spatial form of the Tangshan Jiaotong University

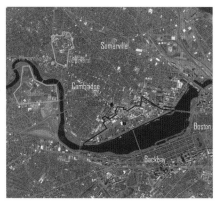

波士顿地区校园空间模式
Campus spatial form in Boston

Harvard: open campus offering urban image

Harvard square   Kresge auditorium

我们的目标是建立三个层级的学校，分别培养具备工匠精神的高技术人才、再教育与才能孵化，以及培养具备多样兴趣的高素质市民。传统的学校是封闭的知识圣殿，我们希望在新的唐山交通大学打造面向城市、活力与生态相依的新高校体系。麻省理工学院与哈佛大学开放的教育空间给我们提供了诸多启示。开放校园提供了别致的城市景致与城市意象。此外，学校设施也向社会开放，更加增强了学校与社会的互动。在这里，借助两大高校激活周边人气，大学面向城市开放，实现活力与文韵并存。我们还在劳伦斯市看到了社区学院，它与周边联排房屋别无二致，但面向周边社区居民开放，是一种几乎零门槛的教育空间。

Our goal is to establish a three-level education system, hoping to cultivate high-tech talents with craftsmanship spirit, to provide re-education and talent incubation and high-quality citizens with diverse interests. The traditional school is gated-school, which is a closed knowledge palace. We hope to create a new university system that is open to the city, being dynamic and ecological. As we traveled to Boston for a city tour and visited the Massachusetts Institute of Technology and Harvard University, their open educational space provided us with many inspirations. First of all, their open campuses provide a chic city view and city image. In addition, university facilities are also open to the public, which further enhances the interaction between the university and the society. The two universities have activated the surrounding popularity. This is the scene we see, being open to the city, and the coexistence of vitality and rhyme. In addition, the community college we saw in Lawrence City looks the same as the surrounding townhouses, but it is open to the surrounding residents, which is almost an educational space with no barriers.

我们对地段上的工业遗产进行评估，筛选出14处重点工业遗产，将其作为工程学院和技术学校的教学场地。根据这14处工业遗产的特性及周边地区的需求，整合设定成7个教育组团，分别为文化艺术传承研究院、唐山交通大学校园、陶瓷技艺学院、陶瓷艺术学院、陶瓷绘画学院、服务技能培训学校以及科技研发展示中心。

By evaluating the industrial heritage on the site, 14 key industrial heritage sites were selected for engineering and technical schools. According to the characteristics of these 14 industrial heritages and the needs of surrounding areas, they are integrated and designated as 7 educational groups, namely, Institute of Culture & Art, Tang-Shan Jiao Tong University (TSCTU) Campus, Ceramic Tec College, Ceramic Art Institute, Ceramic Painting College, Service Skill Training, and R&D Exhibition Center.

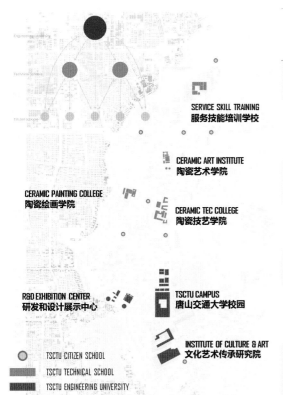

SERVICE SKILL TRAINING
服务技能培训学校

CERAMIC ART INSTITUTE
陶瓷艺术学院

CERAMIC PAINTING COLLEGE
陶瓷绘画学院

CERAMIC TEC COLLEGE
陶瓷技艺学院

R&D EXHIBITION CENTER
研发和设计展示中心

TSCTU CAMPUS
唐山交通大学校园

INSTITUTE OF CULTURE & ART
文化艺术传承研究院

TSCTU CITIZEN SCHOOL

TSCTU TECHNICAL SCHOOL

TSCTU ENGINEERING UNIVERSITY

7 schools	Group oriented	Training object	Investor and Sources of funds	Space size requirement
Institute of Culture and Art	Citizen	Inheritor	Public investment	Larger space
TSCTU Campus	Modern skilled worker	Enterprise production technology backbone	Enterprise	Larger space
Ceramic Tec college	Student	High-skilled artisan	Social funds	Small and medium
Ceramic Art Institute	Student	Mechanic	Social funds	Small and medium
Ceramic painting college	Citizen/student	Interest/artisan	Social funds	Small and medium
Service skill training	Workers from rural areas	Self-employed	Small and medium-sized private Training company	Flexible
R&D Exhibition Center	Worker/intellectual	Enterprise technology innovation talent	Business investment/public subsidy	Scale Large and concentrated

陡河东岸校园功能图
Functions of campuses at the Douhe East

第三级市民学校将之与社区更好地融合、布局与共享使用。回看鲁尔区，其由于教育的存在，多样化的城市社区被建立，而社区也为城市、为教育提供了创新力量。我们按照社区区位均好性进行市民学校的布局，社区学院如豆子般洒入社区中，让每一个市民能够均等地享受到教育带来的城市共享服务。

For third-tier citizen schools, we prefer to integrate them better with the community. Looking back at the Ruhr, due to the existence of education, a diverse urban community was established, providing an innovative force for the city and education. Therefore, according to the uniformity principle, the community schools are sprinkled like beans in the community, so that every citizen can enjoy the sharing service brought by Tangshan Jiaotong University.

○ TSCTU CITIZEN SCHOOL

陡河东岸社区学校分布图
Community schools at the Douhe East

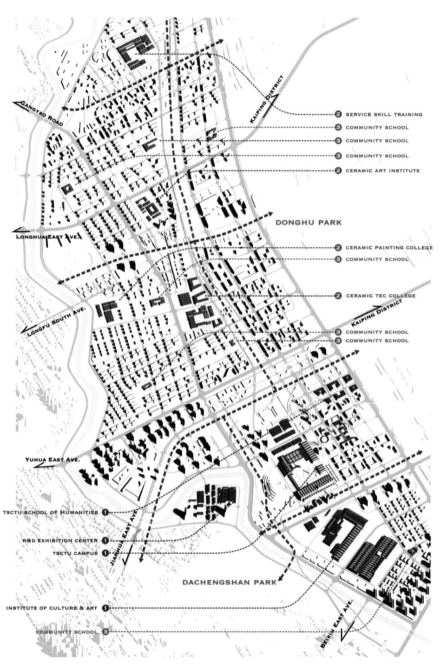

陡河东岸三级学校体系示意图
Three-level education system at the Douhe East

第2步 空间：构建与城市要素的关联
Step 2 Space: linking to the urban elements

考察麻省理工学院与哈佛大学可以发现，开放校园塑造了别致的城市景致与城市意象。在大学周边，诸多科研机构、办公区等园区也应运而生，都证明了开放校园的意义所在。

MIT and Harvard University have given us a lot of inspiration, where open campuses provide a unique urban landscape and image. In addition, many scientific research institutions, office areas and other parks around the city have emerged as the times require, which also proves the significance of open campus.

哈佛与麻省理工案例研究
Cases study of Harvard & MIT

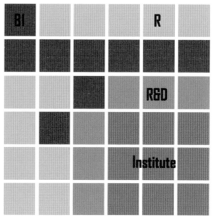

教育导向发展的土地利用模式
Land use pattern of education-led development

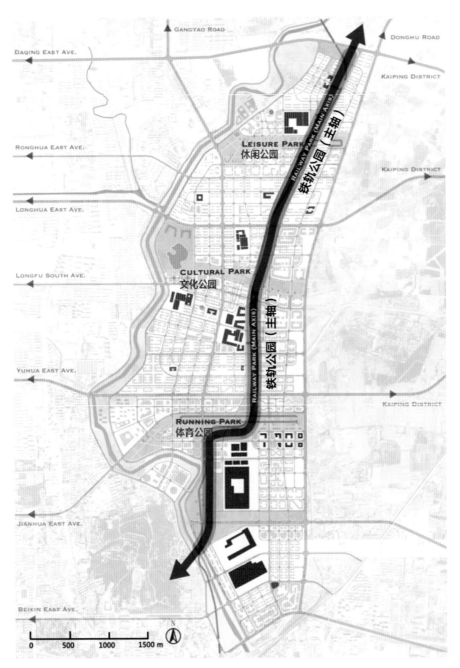

公共空间网络的连接
Link of public space network

首先要构建公共空间网络。陡河东岸工业沿铁轨而生，当工业遗产成为新教育空间时，铁轨成为串接它们的天然线路。因此，散布于地段内的各类校区将被一个铁轨公园完全串接，并为铁轨公园赋予不同的活力。

Firstly, we build the connection of public space network. Since industry along the Douhe East came into being along the rails, when the industrial heritage became the new educational space, the rails became the natural line connecting them. Therefore, separated campus of Tangshan Jiaotong University will be connected by a railway park, as well as providing different vitality to the railway park in turn.

轨道公园与学校系统图
Railway park & school system

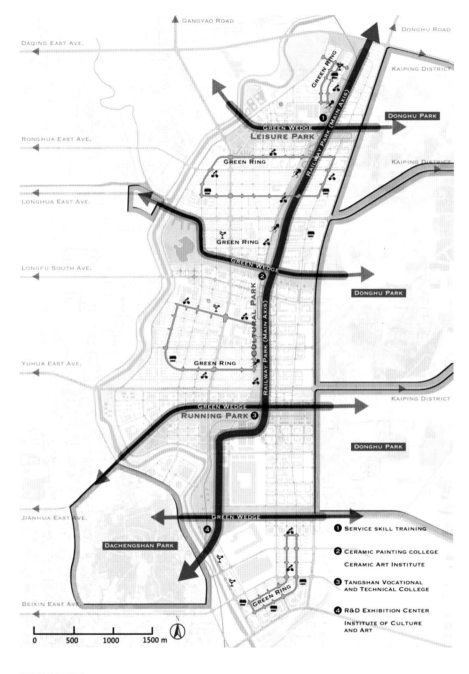

在此基础上，进一步将学校与更大范围的公共空间体系串接。设计方案结合现状绿地，构建了四条横向的绿廊，西接陡河公园，将校园的活力引至陡河，带动沿岸发展；东接东湖公园，将城市与郊野完美融合，提升校园品质。

On this basis, considering linking the campus with a wider public space system, we constructed four horizontal green wedges, connected with Douhe Park in the west, leading the vitality of the campus to the Douhe River to drive the waterfront development, and connected with Donghu Park in the east to integrate into the city and the countryside to improve the quality of the campus.

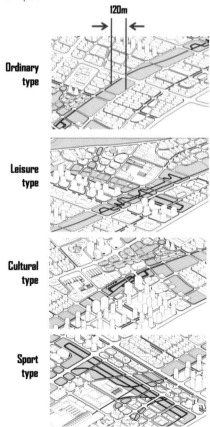

城市绿地系统图
Urban green space system

知识产业网络的连接
Link of knowledge industry network

滨河高端商业
文化博物馆
体验型商业天街
服务技能培训学校
生活技能创业基地

以生活技能为核心的生活服务业节点

文化博物馆
服务技能培训学校
滨河高端商业
主题馆创业基地
体验型商业天街

教育培训类创业基地
陶瓷艺术学校
艺术家创意中心
绘画学校
商业中心
陶瓷技艺学校

以艺术为核心的教育业节点

教育培训类创业基地
陶瓷艺术学校
艺术家创意中心
商业中心
绘画学校

KAIPING DISTRICT

能源类产业孵化
地面商业街区
唐山交大综合校区
唐交人文类学院
科研与技术服务业办公
环境工程学院
新材料产业孵化
文化产业办公
科技会展中心
唐山非遗学校

以新技术为核心的科研节点

其次要构建知识产业网络，让大学发挥最具价值的知识效应。在设定了三个节点后，对节点周边的功能进一步进行了完善，布局更多的科技、孵化、展览、办公商务等空间，以实现区域联动。例如，在以新技术为核心的科研节点，设置有能源类、材料类产业孵化园以及科技会展中心，以期形成良好的"科研-转化"生态链。还设置有科研与技术服务业办公区，以满足大中小企业的办公需求。

Secondly, we build a knowledge industry network to enable the university to play its most valuable knowledge effects. After setting up three nodes, we improve the functions, and implant more science and technology parks, incubation parks, exhibition spaces, office and business areas to achieve functional linkage. For example, in scientific research node with new technology as the core, energy and material industry incubation parks and technology exhibition center are set up to form a good "research-transformation" ecological chain. Also, office areas for research and technology services are set up to address the needs of small, medium, and large enterprises.

地面商业街区（backbay）
KAIPI
唐山交大综合校区
唐交人文类学院
能源类产业孵化
新材料产业孵化
科研与技术服务业办公
环境工程学院
科技会展中心
唐山非遗学校

知识产业网络图
Knowledge industry network

社区生活网络的连接
Link of community life network

我们创建了大学-社区生活网络，让城市、社区与大学形成一个有机体，成为创新的发动机。我们基于现状，梳理基本社区单元，建立社区绿环串接所有的社区学校和社区活动空间。

At the same time, we create a university -community life network, so that cities, communities and the university can form an organism and become the engine of innovation. Therefore, based on the current situation, we first sort out the basic community units, and then establish the green rings of the community to connect all the community schools and community activity spaces.

保留社区与新的社区
Preserved communities & new communities

社区绿环图
Community green rings

创意社区肌理图
Creative community texture

创意街道图
Creative streets map

方案中建立了三个创意社区，使之成为激活唐山创造力的典范。根据三处节点性质的不同，分别设立了创业社区、艺术社区和生态社区，并对其访客来源、投资与维护费用渠道、教育的参与模式等进行了构想与研究。

In addition, we have also established three creative communities to serve as a model for Tangshan's creative life. According to the different natures of the three nodes, we set up entrepreneurial community, art community and ecological community, and conceptualize and study the population source, channels of investment and maintenance cost and educational participation model.

社区类型	人群来源	投资与维护费用渠道	教育参与模式
创业社区	生活服务业创业者	政府投资成立创业社区公司	创业社区公司与学校合作搭建创业平台
	创业团体	创业社区公司通过政府资金和一部分社会融资进行建设	学校为社区创业人员进行培训和资源介绍
	有副业的白领	公司负责为创业者搭建创业平台	社区创业人员通过公司回馈学校红利
	具有技能的创业公司雇员	创业者通过股份红利回馈公司	社区的公司也能为学生提供就业与实习机会
艺术社区	艺术家	校方或陶瓷公司、地产公司出资	学校将一些展览会置入社区
	白领阶层	通过正常的市场机制获利	学校定期在社区内开设讲座与培训
	学校教职人员	通过合约以降低租金、适当降低房价或面积补偿的形式与艺术家达成协议	社区的艺术家可以享受学校的部分资源
	部分普通市民	艺术家对公共空间、城市文化的建设将直接为其创造价值	
	在校学生	政府对艺术家/租户的人口比例设定最低标准	
科技社区	在校学生	地产商或校方出资，政府对地产商拿地地块和容积率做严格限制	校方为其提供专利与技术
	科研人才	建筑或公共空间对于具有技术创新型的设计予以支持	校方将一些公共空间作为实验基地
		根据专利实行情况实施容积率补偿	设置更多的城市实验室、图书馆与市民共享
		设计与实施情况的评定将由多方共同决策	

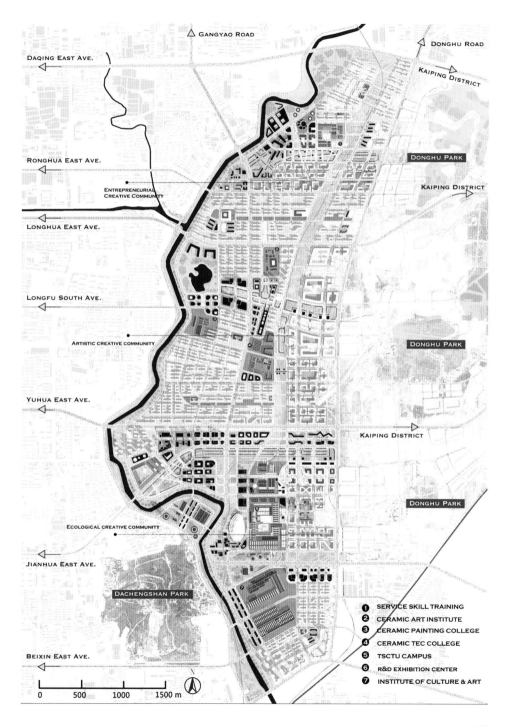

△ GANGYAO ROAD

▷ DONGHU ROAD

KAIPING DISTRICT

◁ DAQING EAST AVE.

DONGHU PARK

◁ RONGHUA EAST AVE.

KAIPING DISTRICT ▷

ENTREPRENEURIAL
CREATIVE COMMUNITY

◁ LONGHUA EAST AVE.

◁ LONGFU SOUTH AVE.

ARTISTIC CREATIVE COMMUNITY

DONGHU PARK

◁ YUHUA EAST AVE.

KAIPING DISTRICT ▷

DONGHU PARK

ECOLOGICAL CREATIVE COMMUNITY

◁ JIANHUA EAST AVE.

DACHENGSHAN PARK

❶ SERVICE SKILL TRAINING
❷ CERAMIC ART INSTITUTE
❸ CERAMIC PAINTING COLLEGE
❹ CERAMIC TEC COLLEGE
❺ TSCTU CAMPUS
❻ R&D EXHIBITION CENTER
❼ INSTITUTE OF CULTURE & ART

◁ BEIXIN EAST AVE.

N

0   500   1000   1500 m

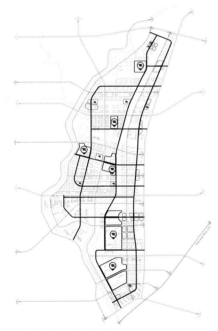

第3步 可达：路网与街道
Step 3 Accessiblity: roads network and life streets

在城市快速通勤方面，为了保护陡河、轨道公园等开放性、慢生活空间，同时契合主干路的基本密度，方案纵向地在陡河与铁轨之间、铁轨与东湖之间规划两条南北贯穿的主干路，横向强调与陡河西岸城市功能的连接。配合陡河东岸不同类型的功能，我们将次干路区分为交通线和社区线两类，前者串联场地上的重点公共功能，在满足交通需求的情况下营造舒适的街道环境；后者是场地上社区交通的骨架，与社区内部绿环紧密结合，营造舒适宜人的街道氛围。最后，我们在道路网骨架上添补支路，设定紧凑宜人的街道高宽比，为行人提供闲适的步行体验。

To ensure the accessibility of the university by main roads so that Tangshan Jiaotong University can play an important role in urban activation and regional radiation, we plan arterial road to establish the supporting system, secondary arterial roads for public services and traffic branch roads for communities. Finally, we add pedestrian sidewalk for high-quality daily life.

Arterial roads network

Branch roads network

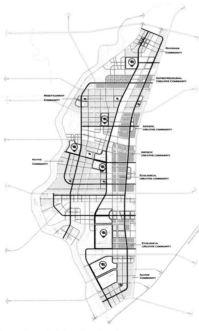

Traffic branch roads network

Secondary arterial roads network

路网与街道
Roads network and life streets

节点深化设计方案
Detailed site plan and programming

通过节点深化设计，可以为新唐山交大三级学校如何与社区、绿地、交通进行融合，提供更加详细的参考。第一个节点是在南部利用唐钢厂区构建的唐山交通大学主校区及其周边的活力住区。正如平面所呈现的，节点设计将大体量的唐钢炼钢车间转换为主教学楼、图书馆与唐交校史博物馆，围绕厂区西侧的贾家山营建一片商业区。同时，由于这个区域有一定程度的土壤污染，我们将所有受到污染的土壤堆积在两个街区里，将之建设成为一片生态修复公园。这个区域中将会复建唐山交大环境工程系，辅助进行该地区的生态修复。同时，部分学校设施面向公众开放，包括图书馆、音乐厅、艺术展廊和创新中心。

The first node is the main campus of Tangshan Jiaotong University with neighboring creative communities. We convert the huge-scale workshop of Tangshan Steel Factory into a main teaching building with a student center, a library and a university history museum. Some amenities are open to public, including libraries, concert halls, art galleries and innovation centers. As for the creative community, we notice that there was soil pollution on that area, so we decide to move polluted soil to the two surrounding blocks and build an ecological restoration park. And we rebuild department of environmental engineering school of Tangshan Jiaotong University to help ecological restoration in this area.

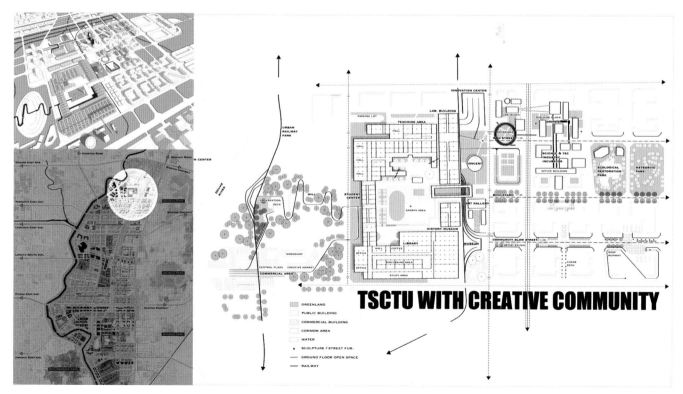

新唐山交大主校区场地设计
Site plan of the main campus of new Tangshan Jiaotong University

第二个节点是坐落在场地中部的陶瓷学院。我们将这里的工业遗产建筑划分为两个校区，包括陶瓷技艺学校和陶瓷艺术学校，同时复建唐山交大工业设计系，提供R&D支撑。我们利用存量空间为学生营造教学、讨论、实习锻炼的空间。整个场地通过艺术街道、商业街道与相邻的创意社区进行积极的联系。此外，西侧有一条廊道实现了两个不同层级的教育机构——陶瓷学院和社区学院的沟通。校内宜人的环境和富有活力的氛围将为大家带来持续的创造力。整个区域也将与邻近的创意社区以及内部艺术街和商业街联系成整体。设计方案预留了视线通廊和景观节点，使得学校的各个位置都会与陡河东岸重要的工业遗产标志建筑形成极好的空间和视线联系。

The second node is the ceramic school located in the middle of our site. We divide the whole ceramic industrial heritage area into 2 faculties, including ceramic pottery skill school in the north and ceramic art school in the south. We set workshop for student internship, meeting room and classroom for teaching and discussion. And this whole area will have a positive link to the neighboring creative communities with all the internal art street and commercial street. What's more, as we can see from the west side, we establish a link between our ceramic school and community school, which are two different levels in the education system.

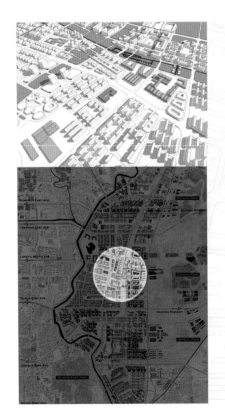

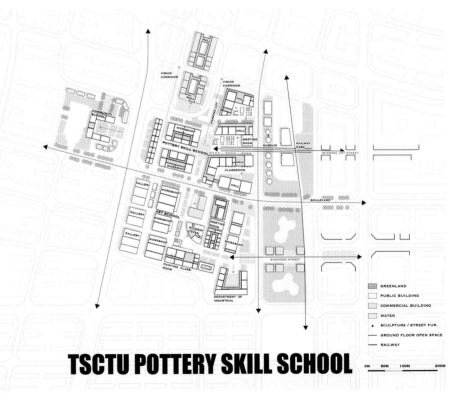

陶瓷学校片区场地设计
Site plan of the ceramic school

TSCTU POTTERY SKILL SCHOOL

第三个节点是北部的服务技能培训学校，它是由唐山第一陶瓷工厂工业遗产改造而来的。由于唐山仍在经历农村人口市民化的过程，带来了大量技术培训需求。在这个地区，我们将植入诸如理发、室内装修、烹饪、城市农场等生活技能培训项目。通过低门槛的生活技能培训，满足步入市民生活的职业需求。我们将在存量空间打造人才中心、实习车间、创客孵化园，为学生提供多种多样便利的学习与职业机遇。面向城市布置开放的繁花步道、运动公园街、都市农场、创意产业孵化工坊，营造积极、开放、绿色的活力界面。

And here comes the node of service skill training school, which is located in the north of our site and is the industrial heritage of former Tangshan No.1 Ceramic Factory. Tangshan is still experiencing the urbanization of the migrant population from surrounding rural area, resulting in a variety of training needs of vocational skills. So in this area, we implant several training programs such as haircut, interior design, cuisine and urban farming. We set up talent center, intern workshop and industry incubator for the students to promote their skills and career opportunities.

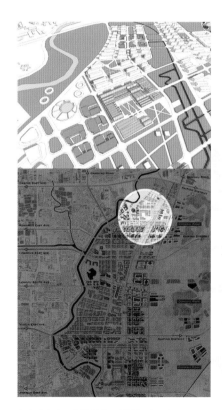

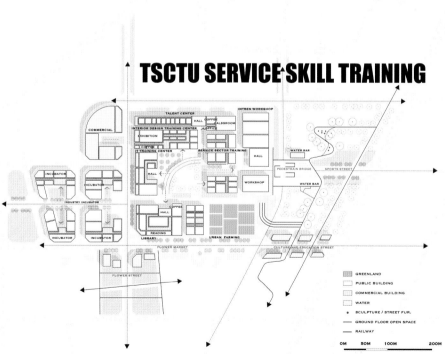

新唐山交大服务技能培训学校场地设计
Site plan of the service skill training school of the new Tangshan Jiaotong University

场地设计1：开放的校园营造了宜人的空间品质。在这里，市民将有机会进行亲子游园体验，感受开放校园的活力。生态修复将带来绿色宜居的自然环境。

Site Plan 1: The open campus will create pleasant spatial quality. Here, citizens will have the opportunity to experience the vitality of the open campus by skateboarding or parent-child tour. Ecological restoration will bring about a green and livable natural environment.

场地设计2：校内宜人的环境和富有活力的氛围将带来持续的创造力。设计预留了视线通廊和景观节点，陡河东岸重要的工业遗产标志建筑在各处均能被观赏和体验。

Site Plan 2: The vibrant campus will bring continuous creativity. As the design reserves sight corridors and landscape nodes, each location of the university has excellent spatial and sight connections with the important industrial heritage landmarks at Douhe East.

场地设计3：营造积极、开放、绿色的活力界面，与城市生活和谐相融。

Site Plan 3: In the design, an active, open, green, and dynamic interface is created to integrate harmoniously with urban life.

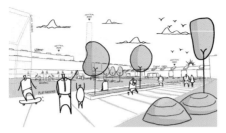

场地设计1 透视图1
Perspective for site plan 1

场地设计1 透视图2
Perspective for site plan 1

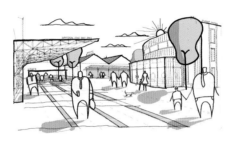

场地设计2 透视图1
Perspective for site plan 2

场地设计2 透视图2
Perspective for site plan 2

场地设计3 透视图1
Perspective for site plan 3

场地设计3 透视图2
Perspective for site plan 3

开发时序及远期目标
Development sequence & long-term goals

终极蓝图不可能一蹴而就，过高的城市负债将会让地方政府面临重大风险，因此陡河东地区的发展转型将会是一个漫长过程，城市居民的当前问题也不可能在短期内解决。制定一个合理的开发时序和开发模式，分批次进行实施，将会有助于这些问题的解决。在陡河东岸，唐山交通大学的大学部分按学院分批建设改造，并在渐进式的过程中完成各项行政审批工作，逐步实现唐山发展、创新的新源泉的建成。

The ultimate blueprint cannot be achieved overnight, and excessive urban liability will expose local governments to significant risks, so the transformation of Douhe East will be a long process, and the current problems faced by the residents cannot be solved at one stroke. The design of a reasonable development timeline and development pattern will help to answer both questions. On the east bank of the Douhe River, we carry out the transformation of Tangshan Jiaotong University in batches of school, and complete various administrative examinations and approval works in the process, to gradually build up the new engine of development and innovation of Tangshan.

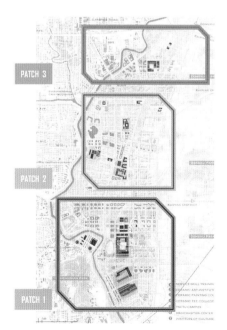

开发分区及时序
Development zoning and sequence

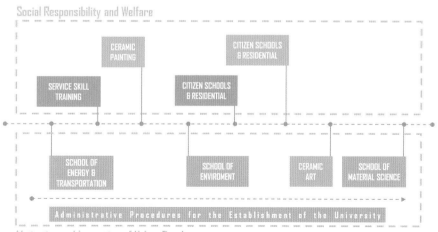

类别	面积 / km²	占比 / %
学校	1.70	12.80
居住	4.47	33.65
商办	1.32	9.90
绿地河流	2.85	21.40
道路	2.96	22.25
总计	13.30	100

学校	教师 / 人	学生 / 人	职工 / 人
清华	3458	48 739	10 000
MIT	1056	11 574	10 000
唐交（以MIT估算）	2083	29 400	10 000
总居住人口（1.8容积率）	总居住人口（1.5容积率）	教育相关人口	创新园区就业人口
201 471	167 892	41 483	35 000

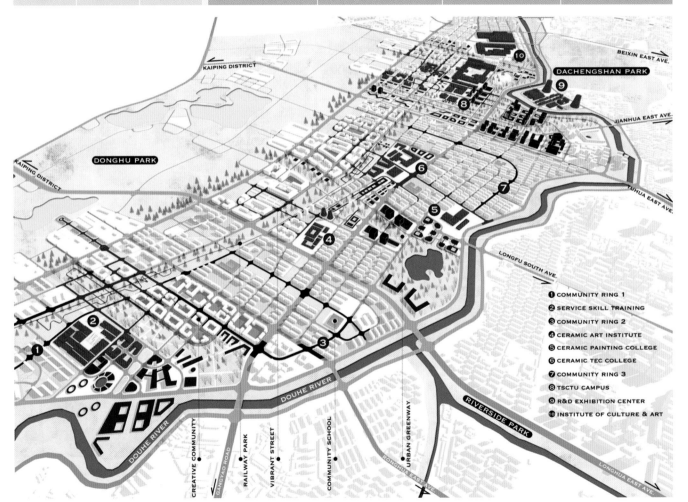

1 COMMUNITY RING 1
2 SERVICE SKILL TRAINING
3 COMMUNITY RING 2
4 CERAMIC ART INSTITUTE
5 CERAMIC PAINTING COLLEGE
6 CERAMIC TEC COLLEGE
7 COMMUNITY RING 3
8 TSCTU CAMPUS
9 R&D EXHIBITION CENTER
10 INSTITUTE OF CULTURE & ART

远期目标及城市设计愿景
Long-term goals and urban design vision

心得体会
Summary

郝恩琦
HAO Enqi

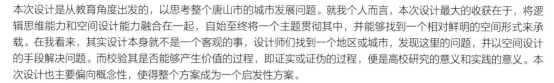

本次设计是从教育角度出发的，以思考整个唐山市的城市发展问题。就我个人而言，本次设计最大的收获在于，将逻辑思维能力和空间设计能力融合在一起，自始至终将一个主题贯彻其中，并能够找到一个相对鲜明的空间形式来承载。在我看来，其实设计本身就不是一个客观的事，设计师们找到一个地区或城市，发现这里的问题，并以空间设计的手段解决问题。而校验其是否能够产生价值的过程，即证实或证伪的过程，便是高校研究的意义和实践的意义。本次设计也主要偏向概念性，使得整个方案成为一个启发性方案。

Starting from the perspective of education, this program probes into the urban development issue of the whole city of Tangshan. Personally, what I have learnt the most in this design is to integrate logical thinking with spatial design and to find a relatively distinct spatial form to deliver one theme from the beginning to the end. In my opinion, design itself is not an objective thing, but a process that designers find an area or a city, discover the problems, and find solutions by means of spatial design. And the process of value verification for the design, i.e., to prove or disprove the design, is just the significance of researches and practices. Basically, this design is still in a conceptual form, hoping to provide inspirations to practice.

孟祥懿
MENG Xiangyi

我们在设计上尝试了一种较为理想化的新风格，通过利用教育来推动城市复兴。我主要负责的是路网骨架、全部住区和北部深化节点的设计。找不同、思源头应该是这次设计最具有文化意义层面的体会。通过在美国的Final Review和城市考察，在整个教学过程中能够清晰地感受到，由于两国历史积累形成的社会体制、发展阶段与规划模式不同，造成两国学术环境的不同、两校育人模式及学生风格的差异。

The design experimented with a new and more idealistic style, trying to promote urban regeneration via education. I was mainly responsible for the design of road network, the residential areas, the detailed design of the northern nodes. Figuring out the differences and the source should be the most meaningful experience of this design in sense of cultural significance. During the final review and the city tour in the US, I could clearly feel throughout the program the differences of the two countries in social institution, development stage, and planning model accumulated in the past, and the differences in terms of academic environment, education model, and styles of students from the two universities.

朱仕达
ZHU Shida

从美国回北京的飞机上，翻看着手机的相册，恍然发现这小半年攒下的照片只能用"极为多元"来形容，随走随拍的街道照片千奇百怪，这个相册几乎就是一个大熔炉。正是这种极为多元的体验让这半年的设计课独一无二。从设计内容上，清华与MIT做出了完全不一样的设计，丰富的差异性反倒是能把两校、两国的思维模式、关注内容的差异完全显露出来。对规划这一专业、学科、行业的探索还远远没有到头，感谢这一学期重新点燃起了我对它不断探索的动力。

On the flight returning to Beijing from the US, I flipped through the photo library on my phone and realized that the photos collected in the past six months are extremely diverse. With a variety of photos taken on the go, the album is almost a melting pot. It is the extremely diverse experience that makes this half-year design class unique. In terms of design content, students from Tsinghua and MIT ending up with completely different design programs, which were not even in the same a discipline. The richness of the differences in turn reveals the differences in thinking patterns and concerns of the two universities and countries, which is probably the biggest harvest of this semester. Thankfully this semester has rekindled my motivation to keep exploring.

从单位大院到社群共同体
From Danwei Compound to Mixed Community

刘艺　张璐　许可
LIU Yi　ZHANG Lu　XU Ke

背景介绍
Background introduction

城市是复杂有机体，而社区是构成城市的基本单元，同时也是居民生活的地区，直接影响着城市和人的发展。唐山市内住区的一大特点就是存在大量的社会主义住宅——单位大院。因此我们从社区入手，研究陡河以东地区的转型发展之路。

If a city is considered as a complex organism, community would be the basic unit of it, and also the area where residents live, so it directly affects both the city and people. One of the main characteristics of Tangshan's residential areas is the existence of a large amount of "socialist housing"—Danwei compound. Therefore, we started from the local communities to study the transformation process of Douhe East area.

We will study it from three levels: city, community, and residents, and from three dimensions: past, present, and future. We try to sustain the advantages of Danwei and discover problems and opportunities, before putting forward the planning and future vision for the "mixed community".

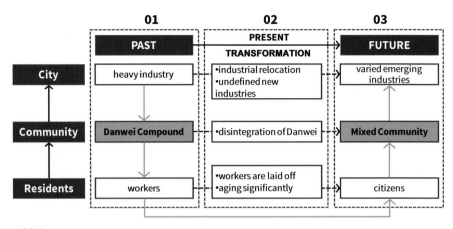

研究框架
Research framework

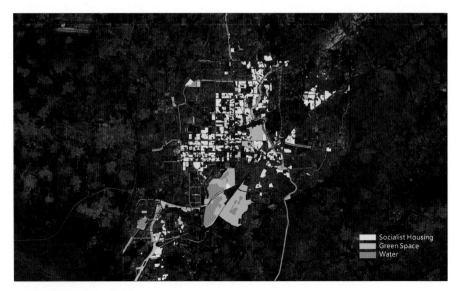

唐山市社会主义住宅分布图
Distribution of socialist housing in Tangshan

唐山单位大院的空间特征：紧密职住关系与紧凑的社区规模
Spatial characteristics: close job-housing relationship & compact community size

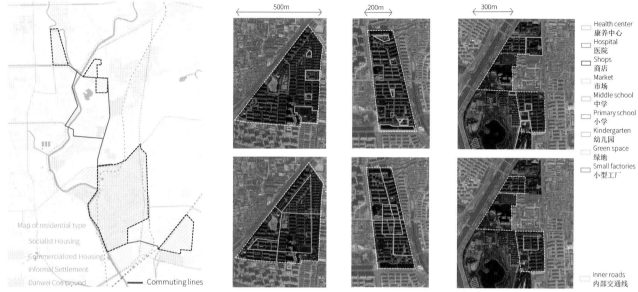

陸河东片区唐钢及其住区研究
Study on Tangshan Steel Plant and its residential areas in Douhe East

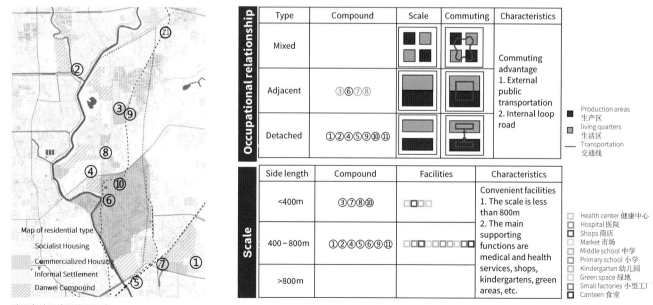

陸河东片区住区类型研究
Study on types of residential areas in Douhe East

唐山单位大院的社会特点：熟人社会关系网与自给自足生活网
Social characteristic : social network of acquaintances and self-sufficient living network

单位大院是唐山社区的主体。陡河东有11个单位大院，他们具有紧密的职居关系与紧凑的社区规模。单位大院具有内部环线或外部通勤设施，配备多种公共服务，生活便利。

Danwei compounds form the main body of Tangshan community. There are 11 compounds in Douhe East, which are featured by a close job-housing relationship and a compact community size. At the same time, they are equipped with either internal loop or external commuting facilities, and are ready with a variety of public services, so the life there is quite convenient.

陡河东中部的电厂工房，是典型的单位大院。这里的北部为生活区，南部为生产区，居民可在15分钟内步行到达任一工作或生活服务设施。生活在单位大院中的工人，日常活动都可以在社区完成。同时，人们靠工作联系在一起，或是工友、或是亲属、或是服务与被服务的关系，形成熟人社会关系网与自给自足的生活网。

The power plant community in the central part of Douhe East is a typical Danwei compound, with the living area in the north and the production area in the south. It only costs 15 minutes for the residents to either go to work or reach a living service facility. The workers living in these areas can complete their daily activities within the community. At the same time, people are connected with each other via work, they are either coworkers, relatives (like fathers and sons), or service providers and receivers, an acquaintances' social network and a self-sufficient living network are thus formed.

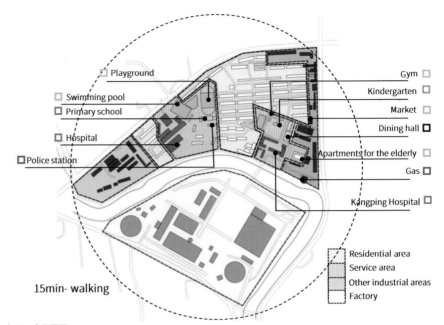

电厂工房平面图
Plan of the power plant community

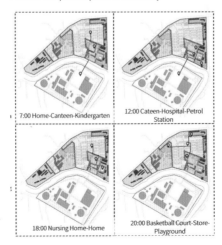

电厂工房一日生活图景
One day life in the power plant community

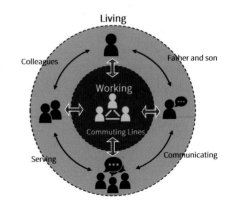

单位大院邻里关系分析
Neighborhood relationship analysis in Danwei compound

096

变化中的唐山：单位制解体
Tangshan in change: dissolution of danwei system

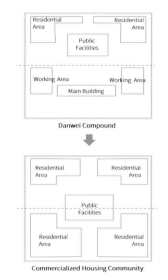

Danwei Compound

Commercialized Housing Community

20世纪90年代，单位制解体，商品房住宅出现。相比于传统单位大院，商品房与工作地点无关，无法享受单位福利，在某些方面无法替代传统的单位大院。

In the 1990s, the Danwei system is dissolved and commercial housing appeared on market. Compared with traditional Danwei compound, commercial housing has no connection with the workplace, no access to Danwei benefits, so in certain aspects they cannot completely take the place of traditional Danwei compound.

从单位大院到后单位社区
From Danwei compound to post-Danwei community

Scenario 1: Danwei remains and new enclosed communities are built up

Scenario 2: Danwei remains, and still retains its neighboring residential area

Scenario 3: Danwei detached, leaving behind housings

陡河东片区住区与单位的三种类型
Three scenarios of residential areas and its Danwei in Douhe East

陡河东片区的现状问题
Current problems of Douhe East

Disrepair

Private construction

Poor market environment

Lack of public space

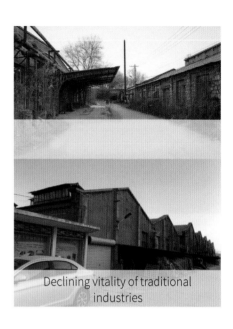

Declining vitality of traditional industries

物质空间问题
Spatial problems

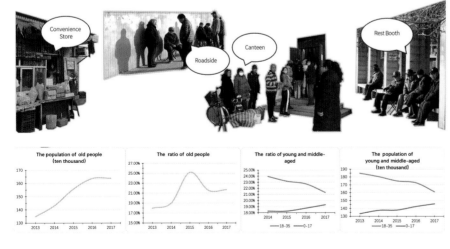

社会问题
Social problems

后单位时代的陡河东地区面临诸多问题：社区缺少管理、废弃工厂未利用、缺少公共空间等，陈旧的物质空间亟待改善。唐山目前人口老龄化、年轻人口外流问题严峻，下岗工人如何就业、退休工人的生活需求如何满足等问题也较为明显。

Douhe East is facing a range of problems at the post-Danwei era: absence of community management, waste of abandoned factories, lack of public space, obsolete material space waiting for renovation, and so on. Besides, other social problems such as aging population, outflow of young people, re-employment of laid-off workers, and livelihood of retired workers are also troubling Tangshan at present.

陆河东片区的发展目标——社群共同体
Development goal of Douhe East: mixed community

2010年后，随着产业外迁曹妃甸，唐山中心城区倡导绿色转型发展，2011年后城市定位转为：生态宜居城市。陆河东需要继承单位大院的优点，同时改善现有问题。陆河东地区东临东湖公园，西临陆河，内有陶瓷公园，靠近大成山，与南湖公园相联系，具有宜人的山水条件；基地内有丰富的工业文化资源与遗产，同时具有单位大院的传统。因此从社区入手，提出陆河东的未来发展目标：构建绿色、宜居、活力的社群共同体。

After 2010, with the industries relocated to Caofeidian, measures were adopted to advocate the transformation to the central urban area of Tangshan and the green development. After 2011, the city development orientation was changed to: an ecological and livable city. Douhe East needs to maintain the advantages of the Danwei compound and address the existing problems at the same time. Considering its pleasant natural conditions, which are peripheral to Donghu Park on the east and Douhe River on the west, with ceramic park inside, and connected with Nanhu Park through Dacheng mountain, and its profound industrial cultural resources and heritage, as well as a Danwei compound tradition, the future development goal of Douhe East is to build a mixed community that is green, livable, and full of vatality.

什么是社群共同体？即通过引入新兴产业，吸引新人群，改善空间环境，重塑社区文化。在提升公共服务、构建水绿网络、更新利用工业遗产的基础上，重塑和谐亲民的邻里关系，创造社群的共同生活与空间共享机会。

How to understand mixed community? It means: by introducing new industries, attracting new residents, improving spatial environment, and remolding community culture, we can not only upgrade public services, build up urban water and greening network, and renew and utilize the industrial heritage, but also re-create a harmonious neighborhood relationship and form a new common life of community and space sharing.

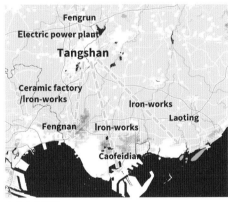

唐山产业外迁
Relocation of industries of Tangshan

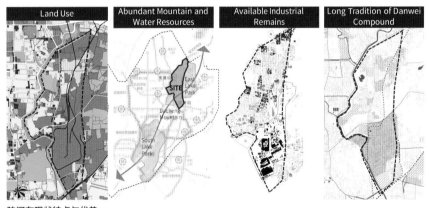

陆河东现状特点与优势
Current characteristics and advantages of Douhe East

Inheriting advantages	Coping with challenges	The future of Douhe East
Convenient facilities Compact community size Convenient transportation Close neighbourliness	Improving infrastructure Refining community management Shaping public space Reusing abandoned factories Meeting the needs of the elderly Attracting young people Reshaping community relations	A greener environment A more livable community A more dynamic industry system 绿色、宜居、活力

未来陆河东片区发展目标
Development goal of Douhe East

设计策略
Planning strategies

产业激活，形成职住平衡的城市组团。振兴产业衰败地区，最重要的是为社群注入新的就业岗位和功能，这样才能够持续地吸引人口，促进职住融合。

Activating the industry and forming urban aggregations with a balanced job-housing relationship. To revive a declined area, the most important thing is to inject new industries and functions into the community, so that it can continuously attract people and promote job-housing integration.

公共空间带动。我们利用公共空间系统，包括路网布置及绿化设置，从空间上建立起基地的骨架，形成多个城市组团，并且帮助建立组团内部的空间结构。

Being driven by the public space. We used the public space system, including the road network layout and the greening system to build up the spatial framework of the base, form a number of urban aggregations, and organizie the spatial structure inside the aggregations.

社群关系重塑。在每个城市组团内部，保留过去单位的优势，形成新的产业混合，帮助构建生活与就业联系紧密的社群共同体。

Reshaping the social relationship. Within each urban aggregation, the advantages of Danwei in the past are retained, and a new mixture of industries is formed, which help build up a mixed community with close connection between life and work.

增强陡河东片区与城市的交通联系
Improve Regional Transportation Linkage

—— Expressway
—— Arterial road
—— Secondary road
—— Branch road

路网结构图
Road network layout

构建完整的城市生态网络
Reconstruct Ecology System

■ Water
▨ Green space

绿地系统图
Green space system

工业遗产再利用
Reuse the industrial relics

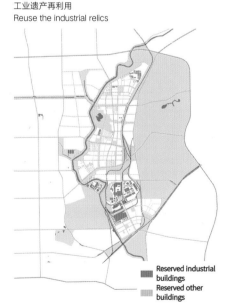

▦ Reserved industrial buildings
▦ Reserved other buildings

保留建筑分布图
Distribution of reserved/reused buildings

构建职住融合的城市组团
Relate Jobs and Residence Tightly

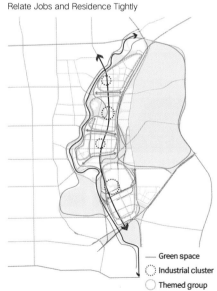

—— Green space
◌ Industrial cluster
○ Themed group

总体结构图
General structure

形成便利高效的交通环
Connected with Public Transit Loop

Medium volume
Public transport
Bicycle/walkway track

公共交通系统图
Public transportation system

提供多种模式的全龄混合住区
Serve for the Whole-Age Population

Long-term lease apartments
Creative community
Mixed-age community
Villa

居住类型分布图
Distribution of various types of residential areas

塑造便利宜居的生活圈
Construct a Mixed Community

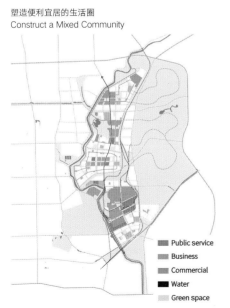

Public service
Business
Commercial
Water
Green space

公共服务设施分布图
Layout of public facilities

增强城市组团功能混合
Functional Fusion

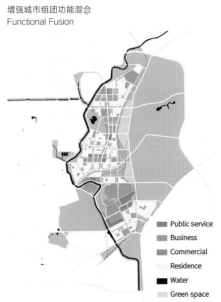

Public service
Business
Commercial
Residence
Water
Green space

总体用地图
General land use layout

在三大策略的指导下，具体的工作重点包括以下三方面：第一，是产业激活，形成职住平衡的城市组团，引入新的产业改善生活、提高城市活力，并对现有的工业遗产与建筑进行评估和保护。

With three primary strategies as the guidance, the specific work focuses include: First of all, it is the industrial activation, the formation of urban aggregations with a balanced job-housing relationship, the introduction of new industries to upgrade life quality and promote urban vitality, and the evaluation and protection of existing industrial heritages and buildings.

第二，改善以及重新设计路网、公共交通、绿色空间骨架，通过公共空间推动城市发展。

Secondly, we improved and redesigned the road network, public transportation system, green space layout, in order to promote urban development through public space construction.

第三，在每个城市组团内部，通过核心产业带动形成就业中心，以公共交通环线为基础形成多个混合社区。混合社区内除居住外，生活功能齐全，其内的公共空间和核心产业有助于居民形成紧密的社会关系。

Thirdly, within each urban aggregation, we tried to form a job center through a core industry, and form a number of mixed communities based on the public transport ring lines. In addition to housing, the living functions in the mixed community are complete. The goal is to help residents establish close social relations through the public space and core industries.

详细设计
Detailed design

我们需要解决的问题包括:

关于城市
基地能在唐山发挥什么作用?
如何将这块场地与唐山的其他地区连接起来?
应该引入或改变哪个行业?
如何利用工业遗迹?

关于社区
单位大院的传统优势是什么?
我们可以从传统的单位模式中学到什么?
我们应该遵循哪些原则?
关于居住模式,如何适应市场需求?

关于居民
居民需要什么?
退休工人将如何生活?
如何建立紧密的居民关系?

The problems we need to solve include:

About the CITY
What role can this base play in Tangshan?
How to connect this base with the other areas of Tangshan?
What industry should be introduced or changed?
How to use the industrial relics?

About the COMMUNITY
What are the advantages of traditional Danwei compound?
What can we learn from traditional Danwei mode?
What principles should we follow?
In terms of the residence mode, how to adapt to the market demand?

About the RESIDENTS
What do the residents need?
How will the retired workers live?
How to build a closely connected neighborhood relationship?

总平面图
Master plan

详细设计：医疗康养组团
Detailed design: health-care community

以北部的医养社区为例展开详细设计，以进一步展示如何实现设计目标。右图展示了基地内不同的功能划分，医养产业集中在西部，交通环线串联起社区内大部分的主要功能。东北部的陶瓷厂被改造为老年大学，最北部设计了一片低密度别墅区，南部及基地内其他地方布置了多种类型的混合住宅。最东部保留一片都市农园与东湖公园衔接，作为城市居民，尤其是老年人的城市花园。

The detailed design is explained with the health-care community on north as an example to further elaborate how to achieve our goals.The figure shows the different functions of the base. The medical and health-care industry is concentrated in the west, and most of the main functions of the community are connected by the transport loop. The ceramic plant in the northeast was reformed into an elderly college, a low-density villa area was designed in the far north, and a variety of mixed-use housing buildings were arranged in the south and other parts of the base. The easternmost part retains an urban farm garden that is connected with Donghu Park, as an urban garden for residents, especially the elderly.

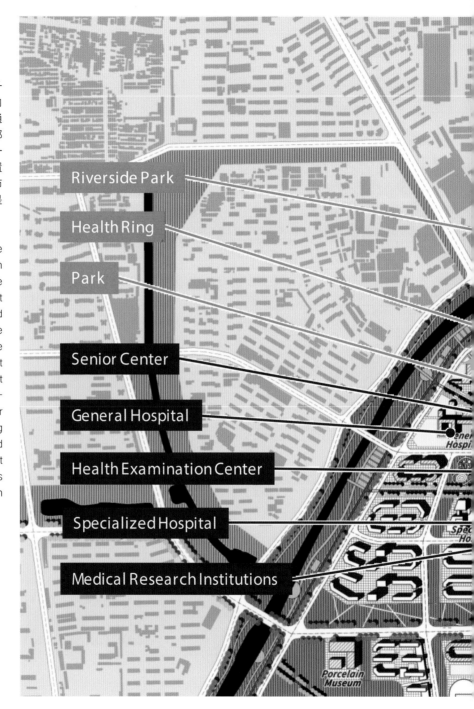

医疗康养组团 总平面图
Master plan of area centering at health-care

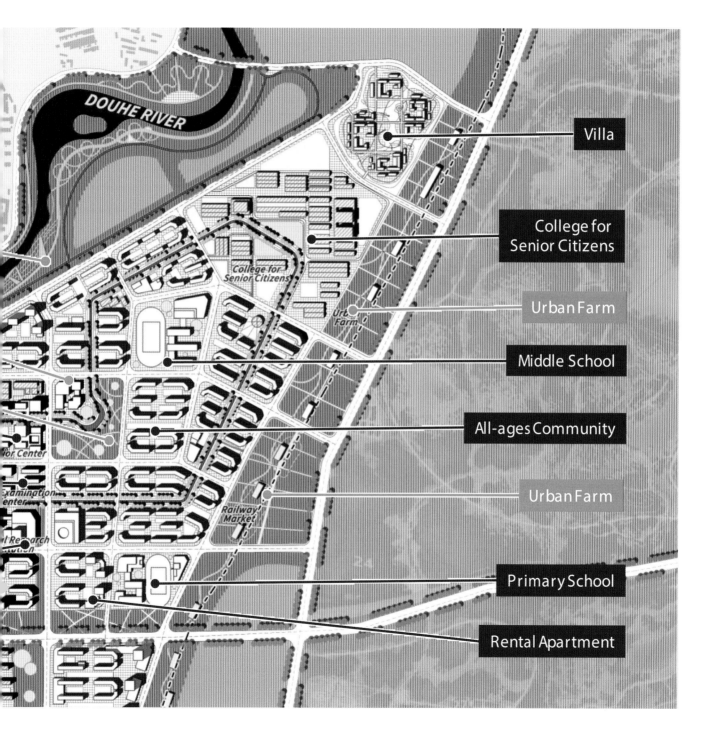

Villa

College for
Senior Citizens

Urban Farm

Middle School

All-ages Community

Urban Farm

Primary School

Rental Apartment

DOUHE RIVER

College for
Senior Citizens

Urban
Farm

Railway
Market

Examination
Center

or Center

Research
tion

健康颐养组团
Health-care community

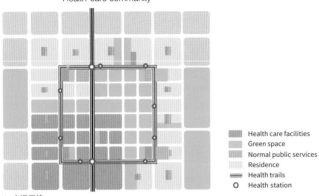

1. 交通系统
Public transportation system
Level 1 Urban public transport corridors
Level 2 Public transit loop: health trails

2. 社区结构
Community structure
Level 1 Concentrated large medical and health facilities
Level 2 Decentralized small recreational facilities

创新创业组团
Innovation & business start-up community

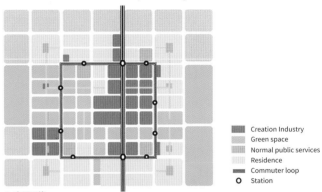

1. 交通系统
Public transportation system
Level 1 Urban public transport corridors
Level 2 Public transit loop: commuter loop

2. 社区结构
Community structure
Level 1 Creativity industries and recreation are distributed along the public loop and go deep into the community
Level 2 Public innovation platform and creative activity area

教育培训组团
Education and training community

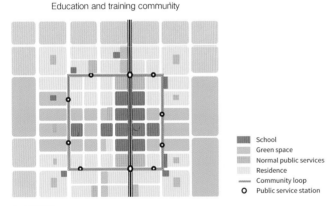

1. 交通系统
Public transportation system
Level 1 Urban public transport corridors
Level 2 Community communication loop

2. 社区结构
Community structure
Level 1 Concentrated education and green space
Level 2 Decentralized all-age education sites

体育运动组团
Sports community

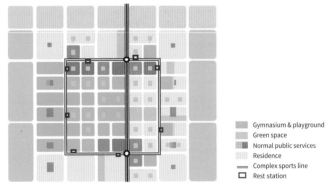

1. 交通系统
Public transportation
Level 1 Urban public transport corridors
Level 2 Public transit loop: complex sports line

2. 社区结构
Community with varies gymnasium
Level 1 Concentrated and complex gymnasium
Level 2 Decentralized and sufficient playground

社区类型
Community type

社区产业集中区，引入了多种模式医疗产业设施，包括综合医院、老年中心等。各设施彼此互补，由社区健康环互联互通，交通十分便利。健康环被植物和景观环绕，内部包含健康步行道、自行车道和老年人交通车道。

A variety of medical industry models are introduced into the industrial concentration area of the community, including general hospital, elderly center, etc. Each model is centralized and at the same time complement each other, which are connected by community health loop, so the transportation is very convenient. Surrounded by plants and landscapes, the health loop contains healthy walking paths, bike paths and traffic lanes for the elderly.

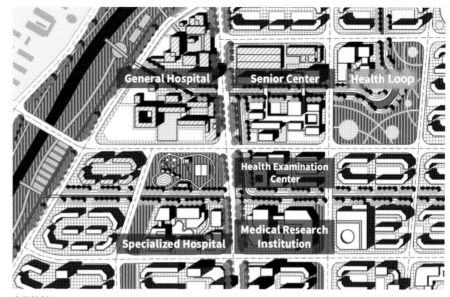

主题策划
Layout design with health-care as the theme

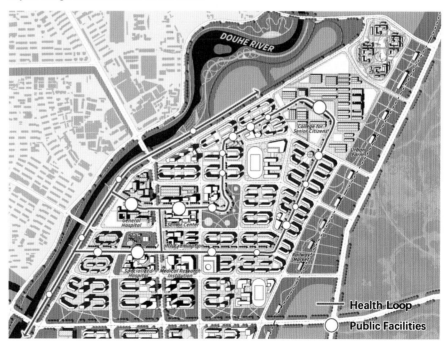

健康环线
Health loop

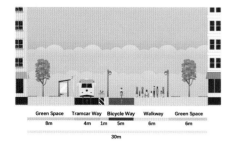

街道剖面
Street section

详细设计：医疗康养组团——工业遗产再利用
Detailed design: health-care community—reuse the industrial heritage

老年大学由原有的大明瓷厂工业遗产改造而来。健康环线从老年大学的内部穿过，将其划分为内外两个区域。通过梳理原有建筑的轴线，形成办公区、教学区和老年公寓三个部分。原有的建筑根据结构分为三种形式进行利用，完整的建筑空间作为教室，屋顶作为半开放的公共空间，桁架作为室外景观。新引入的建筑功能包括了很多对老年人有吸引力的项目，比如园艺、文艺、健康教育等。

The elderly college was transformed from the former industrial heritage of Daming Ceramic Plant. The health loop runs through the inside of the elderly college, dividing it into two parts, inside and outside. By making clear the axis of existing buildings, three parts, i.e., office area, teaching area and elderly apartments are formed. The original building structures are utilized in three forms: the complete building is used as classrooms, the roof is retained as a semi-open public space, and the trusses retained are used as an outdoor landscape.

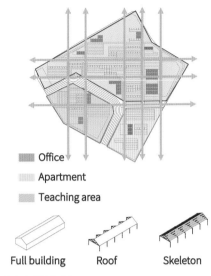

■ Office
▦ Apartment
▦ Teaching area

Full building   Roof   Skeleton

老年大学结构分析
Structural analysis of the elderly college

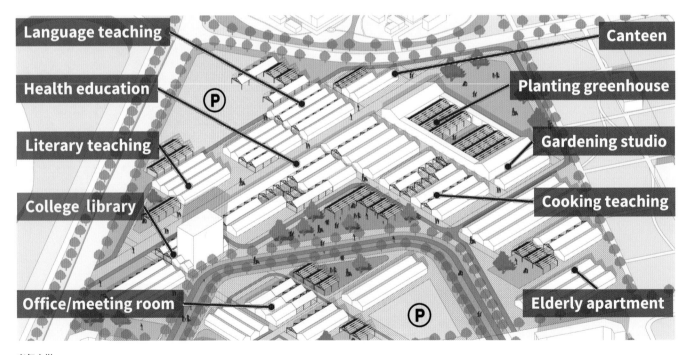

Language teaching
Health education
Literary teaching
College library
Office/meeting room
Canteen
Planting greenhouse
Gardening studio
Cooking teaching
Elderly apartment

老年大学
The elderly college

详细设计：医疗康养组团——老年人的一天
Detailed design: health-care community—one day of the elderly

下图展示了一位老年人一天的生活，和以前单位大院的生活有共通之处。我们可以看到通过混合社区的设计，可以让老年人在一个社区内实现丰富多彩的生活。他们可以得到医疗和养老的服务，在公园内运动、购物，或者去老年大学上学。并且这些活动都被便利的交通环连接着。老年人同样可以通过区域主路与缸窑路前往教育组团，去学习或者接送孩子。各个组团之间通过路网，建立起了相互交织的整体式出行网络。

Figure shows the life of an elderly in one day, which is similar to the life in a Danwei compound before. We can see that through the design of mixed community, the elderly are able to achieve a colorful life within the community. They can get the access to medical and elderly-care services, do exercise in the park, go shopping, or go to class at the elderly college, which are all connected up through a convenient transport line. Moreover, the elderly can also go to the education area via the main road and Gangyao Road to go to class or pick up the children.

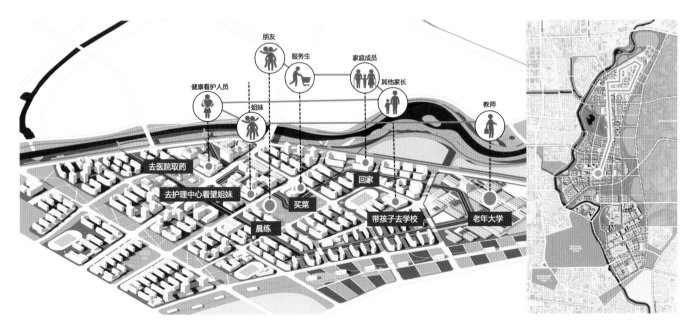

老年人的一天
One day life of the elderly

详细设计：医疗康养组团——年轻人的一天
Detailed design: health-care community—one day of the young people

下图展示了年轻人一天的生活，混合社区不但为他们提供了职住、娱乐功能，还可以使他们便捷地在几个社区之间活动，比如送他们的孩子去教育社区上学，或者去最南部的体育社区参加运动。

Figure shows the daily life of young people. Mixed communities not only provide them with jobs, housing and entertainment, but also make it easy for them to move among several communities for purposes like sending their children to school in education community or participating in sports in the sports community on the south end.

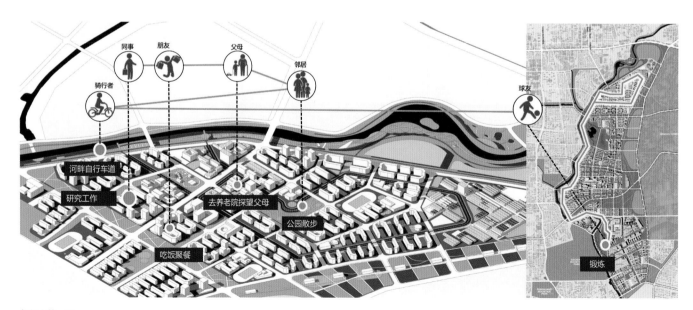

年轻人的一天
One day life of the young people

详细设计：医疗康养组团——季节与场景
Detailed design: health-care community—seasons and scenes

同时，我们会在不同的季节，在社区不同的地方设计不同的景观，如街道公园、步道、铁路绿化带或滨江公园。

At the same time, we designed different landscapes at different locations of the community for different seasons, including street park, walking path, railway green belt or riverside park.

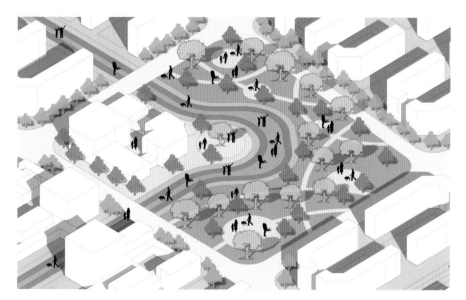

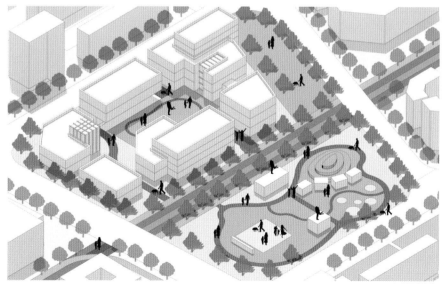

街道公园
Street park

开发时序
Development sequence

在开发时序方面，通过充分考虑现状得出"环境优先—产业带动—配套服务—混合社区"的开发原则，制定"五步走"的开发安排。

In terms of the development order, following the development principle of "priority to environment-industry driven-supporting services-mixed community" that is made based on a full consideration of existing conditions, a five-step development plan was formulated.

Step 1

搭建城市绿色骨架，改善环境；形成道路骨架，增强与城市的联系。
Build up an urban green skeleton and improve the environment; form a road system to enhance connection with the city.

Step 2

引入部分核心产业（医疗康养、教育培训）；对部分住区进行改造提升。
Introduce certain core industries (medical and health care, education and training); renovate and upgrade certain residential areas.

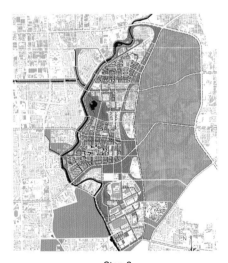

Step 3

配合核心产业，新建混合社区；形成复合公共交通环线。
Support the core industries and build up new mixed communities; form composite public transport ring lines.

Step 4

完善混合社区内部服务功能；推进工业遗产改造，置入新产业（体育、文化等）。
Improve the internal service functions of the mixed community; promote the transformation of industrial heritages and introduce new industries (sports, culture, etc.)

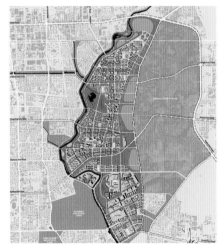

Step 5

增加住宅与配套设施，促进职居融合；预留发展备用地。
Increase residential housing and supporting facilities to promote job-housing integration; reserve development land use.

心得体会
Summary

从1月份在唐山调研，到4月份的线上评图，到5月份在美国波士顿交流，再到6月份回清华汇报，一百多天的时间里，我们在迷糊、纠结、磕磕绊绊中一路走来，但是这并不妨碍大家一起去尝试以更高的视野应对城市发展转型的困境；尝试更开阔的思路，以应对未来时空中的机遇与潜力；尝试更多的挖掘，以应对场地中的人文地域特色；尝试更积极的方式，以应对更有效的协同作业要求；尝试更人性的设计，以回应生活中的种种温情。

From the investigation in Tangshan in January, the online evaluation in April, to the exchange in Boston in May, and then the report at Tsinghua University in June, we have been immersed in a confused, tangled, and stunned feeling for more than 100 days, but it does not hinder us from trying a higher vision to cope with the dilemma of urban development transformation; trying to further open our mind when encounter opportunities and potentials for the future spatial and temporal development; trying deeper exploration of the humanistic and local characteristics; trying more active ways to realize a more efficient collaborative work; and trying more human-oriented designs to reflect the warmth of life.

刘艺
LIU Yi

唐山的历史、人文都揉在了城市里，我们调研的主题是单位大院，陡河东地区内的住区基本都从属于原来的单位。从整齐划一的板楼到小联排，单位的住房各式各样。唐山市民带着有些可爱的唐山口音回答我们的问询，陶瓷工人、唐钢工人、电厂工人都讲述了过去单位的辉煌，流露出对单位消失的一丝惋惜。单位没有了，原有的集体生活消失了，但对于三十几平方米没有客厅的楼房或是别有一番景致的小联排，大家没有太多不满，这大概就是唐山人的知足常乐与惜福吧。这样的单位大院文化深深地影响了我们的方案设计。

Tangshan's history and culture are all intertwined into the city. The theme of our research is the Danwei compound. Subordinate to the original Danwei of local residents, the residential areas in Douhe East are in various forms from neatly arranged slab-type apartment buildings to small-sized townhouses. Tangshan citizens answered our questions with some lovely Tangshan accent. All of them who used to be workers of the ceramic plant, Tangshan Steel Plant, or the power plant told us about the glory of their Danwei in the past, showing slight sentiment for the disappearance of Danwei. Though both Danwei and collective life have gone, they have little complaint about their $30m^2$ large and living room omitted flat or the peculiar small townhouses. This is probably out of the philosophy of contentment and good fortune cherishing of Tangshan people. Such kind of Danwei compound culture has also produced profound influences over our scheme design.

张璐
ZHANG Lu

第一次如此深入地去了解一座城市，唐山是中国众多工业城市中的典型代表，同时又具有其历史意义的特殊性。当我们以研究者、观察者的身份介入这座城市的时候，平日里熟视无睹的现象和事件都变得耐人寻味。当地震发生、企业倒闭、工厂外迁，这些事件都同样在改变着人们生活的空间。在这个过程中，建筑师和规划师也在用自己方式参与着、预测着和追随着。我们

许可
XU Ke

不仅聆听和了解了中国的政府如何在城市发展中发挥作用，在波士顿，我们同样也接触了美国的政府、地产开发商和居民是如何参与城市发展和设计的。这些经历十分宝贵，也十分难得。

For the first time we tried to understand a city in such a thorough way, Tangshan is a typical Chinese industrial city. with its own unique historical significance. When we went into this city as a researcher and observer, the daily phenomena and incidents which we ignore become intriguing. The happening of earthquakes, shutting down of businesses, and moving out of factories all became a part of the force that is shaping the space people live in. In this process, the architects and urban planners are also participating, anticipating and following in their own way. We not only listened and learned about the role of the Chinese government in urban development but also got in touch with the US government, real estate developers, and residents in Boston to understand how they are involved in the urban development and design, which are all very precious experiences to us.

总体鸟瞰图
Birds eye view

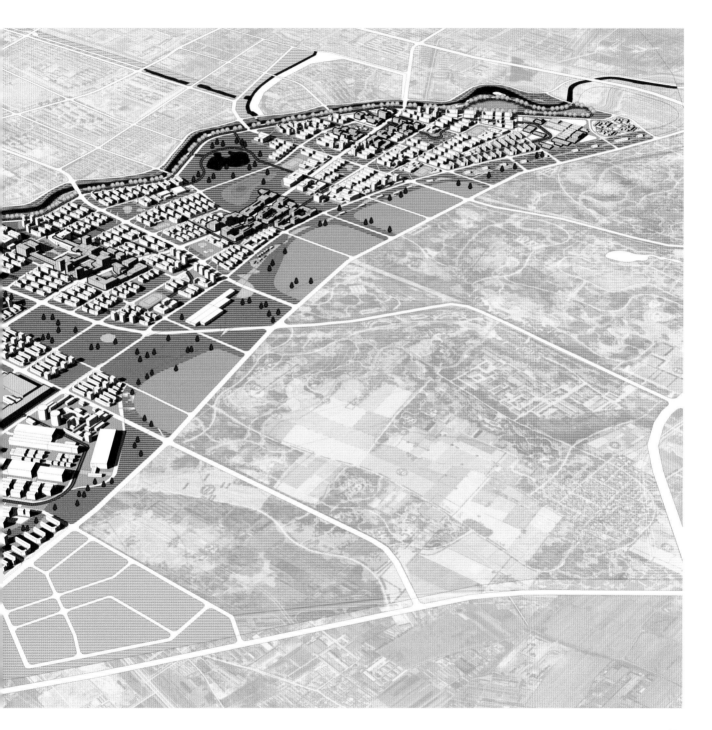

绿于万物
GREEN + X, For a More Livable Post-Industrial
Tangshan

邵旭涛　于旺仔
SHAO Xutao　YU Wangzai

背景介绍
Background introduction

近代的唐山是在洋务运动的推动下，随着开平煤矿的建设而逐渐发展起来的。唐山的城市发展史和工业史密不可分，地段邻近唐山的工业起源地，因此场地内存在大量的材料堆放和工厂遗址。

Driven by the Westernization Movement, Tangshan developed along with the construction of Kaiping coal mines. Its urban development was closely associated with its industrial history. Near the origin of industry, the site is filled with factories and piles of materials.

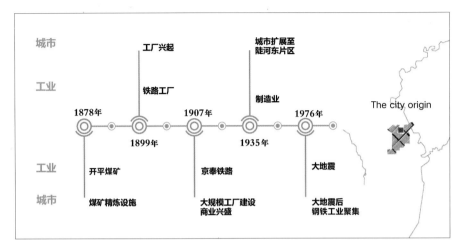

唐山历史与地段关联
Connection between Tangshan's history and the site

研究框架
Research framework

设计通过现状分析推出基于绿色基质的特色解决方案，接下来引进新的产业X，最终产生具体的设计策略。

This design puts forward a green substrate-based solution, then introduces new industry X, and finally brings in a specific design strategy, forming a design scheme of the new Douhe East area.

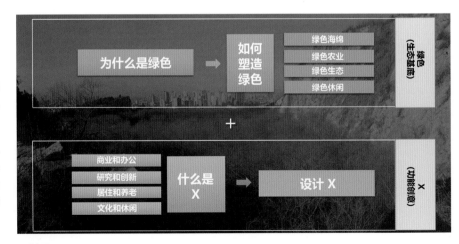

研究框架
Research framework

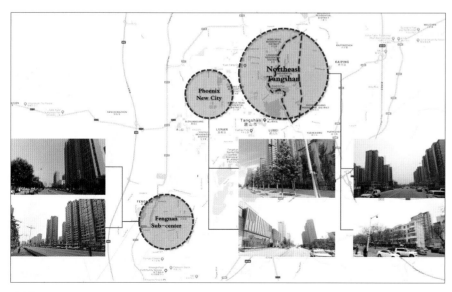

场地问题
Site problems

矿石和煤炭的储存会造成重金属污染，对农作物和人类造成危害。场地内的造纸厂、陶瓷厂和钢厂也会导致有机物污染和土壤酸化等问题。

The storage of ore and coal may cause heavy metal pollution, which is hazardous to crops and human health. Paper mills and ceramic and steel factories can also lead to ecological problems including toxic organic pollutants and soil acidification.

在我们的场地调研中，开放空间的不合理利用让人深思。公园和海滨小道空无一人，而一些街道和一些破败且隐蔽的角落却有市民活动。相比已经实现生态转型的南湖片区和丰南副中心、凤凰新城等，唐山东北部片区主要为老旧小区与零散街角绿地等低品质空间。

Our survey finds that the misuse of open space is impressive. People are hardly found in the parks and waterfront trails, whereas public activities are held in some dilapidated and inaccessible places. Compared with the Nanhu area, Fengnan Sub-center, and Phoenix New City, which have already achieved ecological transformation, the northeastern part of Tangshan features low-quality landscape and old communities.

BaP
(Coking plant)

TOXIC
有毒

Cyanide and organics
(Paper mill &
ceramic factories)

pH value
acidification
(scouring)

场地污染
Site pollution

地段低品质景观
Low-quality open space

唐山的后工业转型期正面临着多重的挑战。近五年唐山的老年人口比例以近1%的增速逐年上涨。车辆制造、钢铁、陶瓷等唐山市区曾经的工业支柱已经或正在搬迁到郊区区县。中心城区的三产占比虽是唐山各区县中较高的,但人均GDP却是倒数。三产品质的情况不容乐观。

Tangshan City is facing many challenges in the post-industrial period. Its aging rate increased by nearly one percent per year in recent five years. Industrial pillars such as steel and ceramics have been or are being relocated to remote counties. Compared with other counties and districts in Tangshan, its inner city has the highest proportion of the tertiary industry yet the lowest per capita GDP.

Aging rate in 2017

Proportion (%)

2013-2017 Aging rate flow

Quantity (Million) / Proportion (%)

☐ Over 60 years old (Million) ── proportion (%)

GDP per capita

GDP per capita (RMB)

Industrial structure

Proportion (%)

☐ Primary industry ☐ Secondary industry ☐ Tertiary industry

唐山的老龄化趋势
Aging rate in Tangshan

唐山各区产业情况(2017年)
Industries in Tangshan (2017)

案例研究
Case analysis

研究世界上其他后工业城市转型发展的方式可发现:鲁尔区的转型是从开放空间的改造开始的。点缀其中的工业遗产反过来又给开放空间系统带来新的推动力。被誉为全美宜居城市的匹兹堡,它转型的第一步是环境治理和基础设施建设,然后才进行大规模的功能改造以及追求更高层次的空间艺术化、商业化。

Research on post-industrial cities in the world finds that the transformation of the Ruhr area began with the restoration of open space, and the industrial heritage in turn brought new impetus to the open space system. Similarly, Pittsburgh started with environmental governance, followed by functional revival.

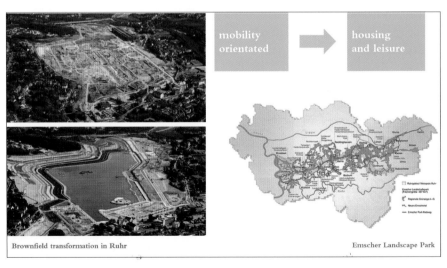

Brownfield transformation in Ruhr

mobility orientated → housing and leisure

Emscher Landscape Park

鲁尔地区以绿色更新推动城市复兴
Urban revival driven by green renewal in Ruhr Region
Source: Ruhr official website

City around Beijing	Main Industry
Tianjin	Technology Finance
Chengde	Agriculture Tourism
Qinhuangdao	Logistics Technology Tourism
Baoding	Technology Light industry
Langfang	Light industry Service
Tangshan	Should consider it's main function to show the advantage

京津冀城市群现状主要功能
Current function of the Beijing-Tianjin-Hebei urban agglomeration

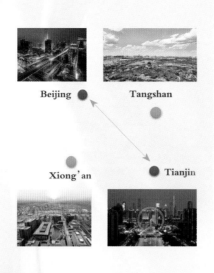

雄安与唐山区位
Location of Xiong' an and Tangshan

外部动力
External drive

外部动力也促使唐山绿化有根本改善。京津冀城市群除了北部山区张家口与承德二市，区域中的城市分工多以科技、轻工业为主。因此唐山作为省级中心城市，应配合新设立的雄安新区承担京津冀两翼差异化的核心功能。

The external drive also urges Tangshan to fundamentally improve its green landscape. The Beijing-Tianjin-Hebei (Jing-Jin-Ji) urban agglomeration needs a better functional division of cities. As a provincial central city, Tangshan should cooperate with the newly set up Xiong'an New Area, to carry the functions of the two wings of the development axis.

方案概念
Project concept

在改善生活质量的内在需求和城市群转型的外部驱动下，我们的设计出发点是创建一个绿色基质城市。

Under the internal demand for a better living quality and the external drive for transformation of urban agglomeration, our design aims to create a green substrate-based city.

绿色基质概念引出
Concept of the project: green substrate

绿色基质设计
Shaping the green

遍布的废弃铁路是地段的特色之一，将其与地形信息叠加，可以观察到，低洼地带和铁路有明显的重合关系。另外铁路附近的防护绿地本身就有一定的绿化基础。因此，将铁路作为地段设计中的绿色骨架，引导建筑功能置入与人的活动。

Abandoned railways are one of the main characteristics of the site. Obviously, railways overlap the low-lying area which is suitable for water collection.

第一步，在地段的低洼区中设计以水景为中心，具有集水功能的公园，实现海绵城市建设目标。

First, we design sunken parks in the low-lying area to meet the requirements of sponge city.

第二步，在地段居民区较多的地方设置了带形的都市农园。在提供另一种城市景观的同时，将不耐运输的花卉、绿叶蔬菜等新鲜作物带到市民生活中。在日本的经验中，高效率的都市农园利用不足2%的全国农业用地创造了8%的农业总产值。废弃铁路的好处是可以将物品运输到地段的各个区域。因此，我们将其打造为都市农场线。滑块和模块化的植物种植区可以提供高效、全季节的活跃农园，满足老年人对种植的热爱，也为儿童提供自然课堂。

Second, urban farms are set up near the residential area. While providing urban landscape, the farms can also bring flowers and green leafy vegetables which are not resistant to transportation to citizens. As the railways can reach everywhere within the site, we built it into an urban farm line.

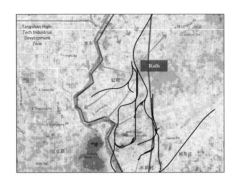

铁路与地形
Topography & railways

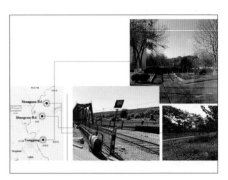

铁路两侧绿地
Existing green space along railways

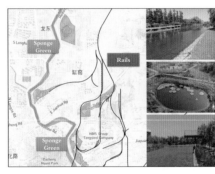

海绵功能绿地
Sponge green space

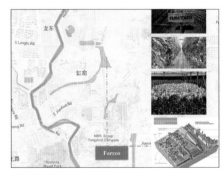

都市农园的位置
Location of urban farm

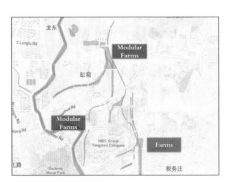

都市农园与模块种植区
Urban farm & modular farms

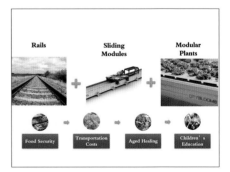

模块概念
Concept of slider farm line

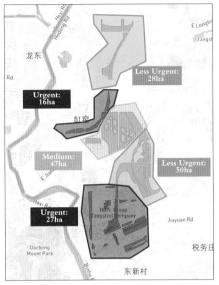

修复时序
Timing of restoration

住区土壤修复
Soil restoration of the residential area

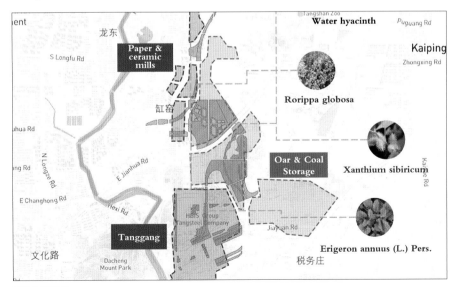

绿地生态修复
Ecological restoration of green space

第三步是作为设计重点的绿地生态修复。在现状煤堆上方及工厂周围种植多重排污植物。场地内整理出168公顷的集中修复绿地。其中位于都市农园、南部综合体共43公顷的绿地需要在其他建设前修复完成，采用植物萃取修复与挥发修复。中部47公顷体育公园景观性强，建设中同时修复。其他78公顷可根据建设强度分期实现，可采用较慢的植物固定修复。

The third step is ecological restoration. Sewage vegetation is planned to grow over the existing coal piles. A total of 168ha of land is scheduled as centralized ecological restoration space. Among them, 43 ha of land is expected to complete restoration before construction, including urban farmland and the complex in southern part. The 47 ha sports park in the central area should be constructed and restored simultaneously. Others can be arranged by stages according to construction density.

位于北部轻工业区与东部住宅周边共46公顷的土地需要修复以达到居住指标。位于南部唐钢厂区的居住区，需先评估其生态污染程度，判断是否需要种植生态修复植物。每年进行一次植物收割和播种。在酸洗池与造纸厂等土壤酸化的地段布置狼尾草。

In addition, a total of 46 ha of land should be restored, to meet the living standards. Heavy metals and toxic organic compounds in soil can be absorbed by different plants sown once a year. Pennisetum plants are arranged in the areas where the soil is partially acidified, such as pickling ponds and paper mills.

与鲁尔区的策略类似，我们将地段的低洼处做浅水塘景观改造，充分发挥一年生水生植物的快速吸附和净化功能，实现生态修复的同时，顾及到景观观赏需求与视觉颜色配比。为了减少动物体内的重金属富集，尽量选用不容易被动物取食的品种。此外，在生态修复公园周边设置硬地铺装，在一部分重点区域设置加密防护网，减少动物的流动。

Some low-lying area is transformed into shallow water pool, giving full play to fast adsorption and purification of annual aquatic plants. In order to reduce the accumulation of heavy metals in animals, hard pavement is set around the park so that plants are not eaten by animals.

工厂周围布置了攀岩壁、浅水池等工业景观。地段内的修复区形成水净化/土壤净化两个生态修复层，修复成功的开放公园的中间层和工业改造项目的活力层，形成生态修复的全层级景观系统。渤海中部沿岸区域为华北重要的工业基地，形成了特殊的景观，也存在大量需要修复的土地。陡河东地段与其他工业遗产集中的生态修复区合作，形成古冶—唐山—丰南—宁河—天津的区域性工业与自然景观公园带，打造纪念性和生态型并重的区域级巨型景观博物馆。

Industrial landscape including climbing wall will be arranged around the factory. The central coastal area of the Bohai Sea is an important industrial base in north China, forming a special landscape, and there is also a large area of land to be restored. The Douhe East area and other zones constitute a regional industrial and natural landscape park belt.

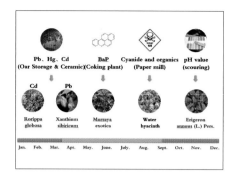

植物选择
Plant selection

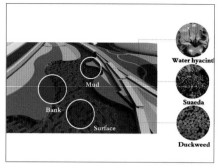

景观与颜色
Color landscape

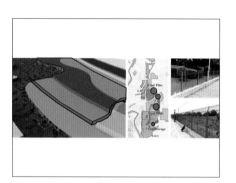

动物隔离
Animal isolation

完整修复景观层级
Complete ecological layer

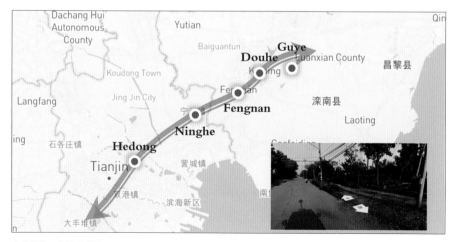

未来渤海工业景观公园
The future Bohai industrial landscape park system

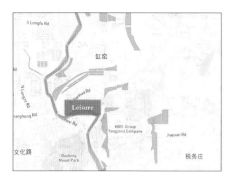

休闲公园
Leisure parks

第四步，通过添加普通休闲公园，对片区绿色网络进行联通。形成海绵功能、生态净化、都市农园、休闲公园四者结合的绿色基质，并使大城山、东湖形成东西向的关联，使陡河西侧城市界面形成绿色连接。

The fourth step is to build ordinary leisure parks, maximizing the connectivity of the green network.Thus, a green substrate combining sponge function, ecological restoration, urban farm, and leisure park is formed. It connects the Dacheng Mountain and the Donghu Lake in the East-West direction. In addition, it also connects the urban interface on the west side of the Douhe River.

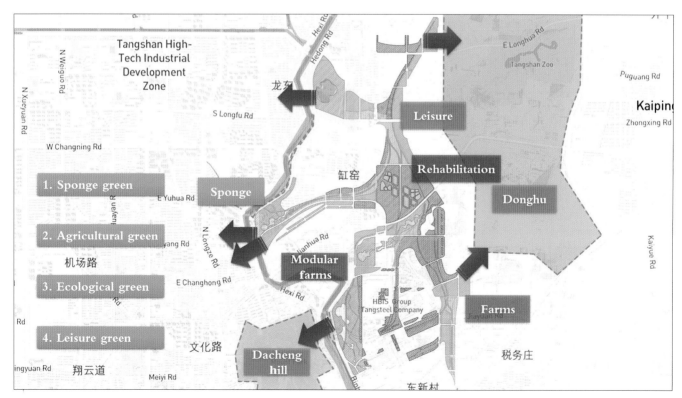

绿色基质总图
Green substrate

现状功能分析与功能业态引入
Introducing the X

X1——商业办公
今天，唐山的经济驱动力已经逐步从投资转向消费，但消费水平仍然相对较低，且城市仍然缺少高品质的消费空间。

X1—Commerce & office
Currently, Tangshan is witnessing a shift of economic driving force from investment to consumption. However, it is still at a relatively low consumption level and lacks high-quality consumer places.

X2——科技创新
从2015年的产业数据可以看出，唐山的第二产业占GDP比例较高，在退二进三的过程中，其科技发展水平低于同级其他城市。

X2—Research & innovation
In Tangshan, the development of industry-related services such as scientific research falls behind that of the secondary industry. The city's expenditure on research and innovation is lower than that of other cities in the same class.

X3——文化创意
唐山以工业而兴，创造了我国工业史上诸多个第一，留下了丰富而宝贵的工业遗产。这些工业遗产伴随着老一辈唐山人的出生、成长、工作和衰老，融入他们生命，成为不可分割的一部分，也是唐山一个重要的对外名片。

X3—Culture & creation
Tangshan's splendid industrial history has left a valuable industrial heritage. These industrial heritage sites are indispensable name cards of the city, accompanying the whole life of the older generation of Tangshan people.

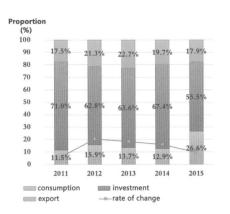

消费、投资、出口占GDP比重
Propotion of consumption, investment and export

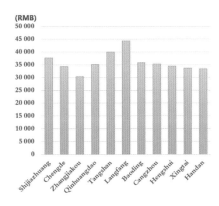

不同城市居民人均年收入
Average wages in different cities

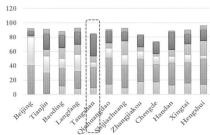

不同城市科技制造水平比较
Comparison of technology-intensive manufacturing industry in different cities

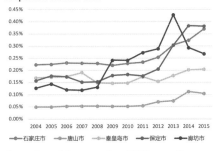

不同城市科研人员和经费的比较
Comparison of researchers and R&D expenditure in different cities

| 1876 | 1878 | 1881 | 1889 | 1896 | 1910 | 1914 | 1919 | 1920 | 1921 | 1937 | 1941 | 1949 | 1954 | 1976 | 2005 | 2016 |

- 1876 唐廷枢组织测绘京唐大运河
- 1878 开平矿务局 揭开中国近代机械平煤的序幕
- 1881 唐胥铁路 中国第一条标准铁路，带动机车发展
- 1889 唐山细绵土厂 中国第一桶水泥
- 1896 北洋铁路官学院 中国第二所高等工科院校
- 1910 华记发电厂 唐山第一座发电厂
- 1914 唐山建筑陶瓷厂
- 1919 华新纺织厂 4年后全部使用机器生产
- 1920 唐山启新洋灰公司 城市功能日趋完善，辐射作用增强
- 1921 中国首台烘干机
- 1937 日本全面发动侵华战争
- 1941 唐山制钢所 为唐山工业增加动力和物质基础
- 1949 中国人民共和国成立
- 1954 启新洋灰公司实现公私合营 工人阶级做主人
- 1976 唐山大地震
- 2005 曹妃甸区成为国家第一批发展循环经济试点产业园区
- 2016 曹妃甸列入国家「十一五」发展规划

第一座机械化煤矿 | 第一条标准轨距铁路 | 第一台蒸汽机车 | 第一桶机制水泥 | 第一件卫生陶瓷 | 第一支股票 | 第一个大学教授

唐山的工业文化历史
Tangshan's industrial culture history

功能业态引入
Introducing the X

Beijing Tianjin Tangshan Qinhuangdao Jinzhou Shenyang Changchun Harbin

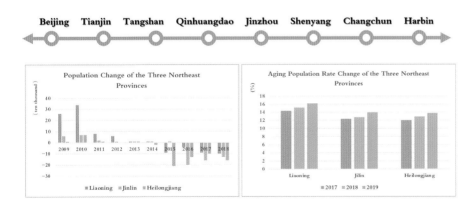

Population Change of the Three Northeast Provinces
（ ten thousand ）
2009 2010 2011 2012 2013 2014 2015 2016 2017 2018
■Liaoning ■Jinlin ■Heilongjiang

Aging Population Rate Change of the Three Northeast Provinces
（%）
Liaoning Jilin Heilongjiang
■2017 ■2018 ■2019

东北的老龄化与人口外迁趋势
The trend of the aging of population and outward movement of population in northeast China

X4——居住养老
唐山位于华北—东北咽喉，在东北的老龄化与人口外迁趋势下，可配合秦皇岛接纳周边地区的老年人口，打造休闲养老之城。

X4—Residence & senior-care
As an important place connecting north and northeast China, Tangshan can serve as a large area for seniors and migrants in the context of the aging of population and outward movement of population in northeast China.

方案生成过程
Planning process

场地要素主要包括几片绿地、工厂群、老旧住宅和大量原材料与煤的堆放地。我们首先拆掉了质量低、难以利用的住宅和工厂。根据唐山城市发展特点，将地段大致分为北部静区和南部活力区。

The existing site elements mainly include green spaces, factory clusters, and large piles of raw materials. We demolished poor-quality residences and factories. According to the layout of other functional areas in Tangshan, this site is roughly divided into two parts: the quiet north area and the active south area. The road network is developed based on the existing roads in Tangshan.

根据地段元素走向和唐山道路已有的关系确定路网。一条绿色自行车道串联滨河公园、大城山、南部功能区和所有住区，既是次级绿廊，也在局部地段作为次级路网，鼓励绿色慢行交通方式。

A green bicycle lane that connects the riverfront, the Dacheng Mountain, the southern functional area, and all the settlements is both a secondary green system and a part of the road network, encouraging non-motorized traffic in the area.

地段原状
Original condition of the site

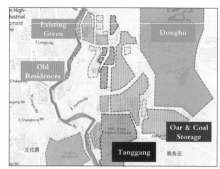

拆除部分住宅和工厂后
The site after demolishing some houses and factories

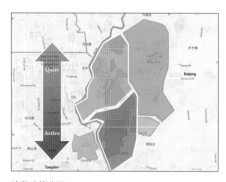

地段动静分区
Division between active and quiet areas

路网生成
Road network

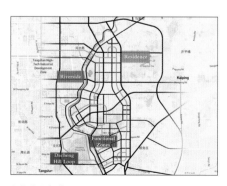

绿色自行车道
Green bicycle lane

用地图
Land-use map

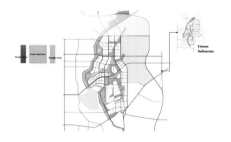

绿色基质
Green-substrate

U形带状住区
U-shaped residence

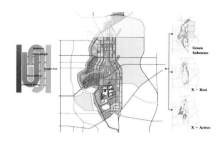

U形活力产业
U-shaped functional areas

双U形结构功能布局
Double U-shaped master plan

引进的新功能X在空间上形成双U形结构。X需要与网状的绿色基质、东湖公园和陡河界面形成最大的接触。在场地的北侧与东侧布置U形带状住区，南部布置活力产业，让活力产业功能区与西侧城市界面形成较好的关联性，住区最大限度接触东湖的自然景观。通过打造绿色基质与双U形功能区交叉的形态结构，形成地段的总平面。

The new function X has a double U-shaped structure in space. X needs to have the maximum contact with the green substrate, the Donghu Park, and the Douhe river interface. The functional areas should be better correlated with the urban interface. These systems finally form a master plan for the site.

X1——商业办公
X1—Commerce & office

X1是绿+商业办公。功能板块是位于陡河沿岸的唐钢旧址，利用遗留的工业建筑遗产景观创造宜人舒适的空间，吸引人们到这里工作和消费。

X1 is Green+commerce & office. It's located at the original site of Tangshan Steel Plant on the riverside. We aim to create a pleasant and comfortable space using rich industrial architectural landscape, to attract people to work and consume there.

保存下来的工业建筑、改造的商业综合体可以为游客提供公共绿地以令其充分享受工业景观。在陡河沿岸新建设的办公建筑则面向唐山电厂和大城山，有着极佳的景观视野。

The preserved industrial buildings and transformed commercial complexes, are providing visitors with public green space to enjoy the industrial landscape. The newly built offices by the Douhe River are facing the Tangshan Power Plant and the Dacheng Mountain. From the offices, we can capture excellent landscape views.

在政策上，唐山需要争取国家政策，不断扩大对外开放力度，例如设立京津冀自由贸易区、跨境电子商务综合实验区等。

In terms of policies, it is necessary to further open Tangshan to the outside world in line with national policies, such as the Beijing-Tianjin-Hebei Free Trade Zone and the cross-border e-commerce comprehensive pilot zone.

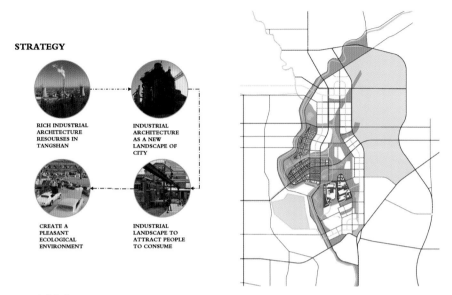

X1——商业办公
X1—Commerce & office

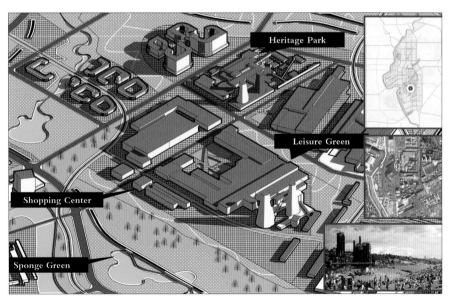

工业建筑遗产改造成为商业空间
Shopping center transformed from preserved industrial buildings

STRATEGY

INDUSTRIAL ARCHITECTURE AS A NEW LANDSCAPE OF CITY

ISSUES//
–INCREASED ABANDONED BUILDINGS
–LIMITED LAND RESOURSE

SOLUTION//
–TO REDISCOVER THE EXISTING STRUCTURE
–ECOLOGICAL TRANSITION

APPLYING INTO THE CITY >>
SYSTEM OF GREEN INDUSTRIAL HERITAGE

ABONDEND STRUCTURE AS A FRAMEWORK FOR
RESEARCH CENTER AND DESIGN INSTITUTE

Incubator

Design Institute& Research Center

X2——科技创新
X2—Research & innovation

X2——科技创新
X2—Research & innovation

X2是绿+科技创新，我们利用废旧的工业厂房建筑，保留其房屋结构框架，改建利用为科技研发中心和设计机构。

X2 is Green+research & innovation. We use abandoned factory building structure as a framework for research center and design institute.

以化整为零的策略，我们将巨大的厂房体量打碎利用，创造更宜人的尺度和更亲切的步行空间。

We break the huge factory building into small pieces and renovate them, creating a pleasant and friendly pedestrian space, and introducing green into the building.

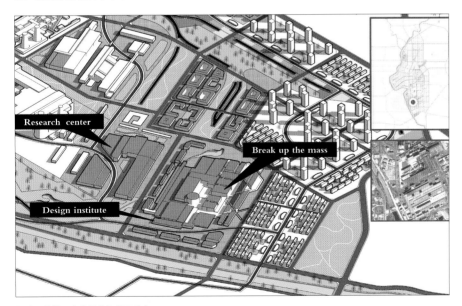

Research center

Break up the mass

Design institute

打碎巨大的厂房体量作为科研中心
R & D center by breaking the huge factory building into small pieces

利用太阳能、垂直绿化和雨水收集等技术构建
绿色建筑系统。多种类的植物种植创造多层次
的公共空间绿化和屋顶绿化体系。科技创新政
策方面，唐山可以学习中关村的系列政策或设
立中关村分园以吸引高新企业入驻，通过校企
合作共享技术，实现双赢。

We develop a green building system using
technologies including solar power, vertical
greening, and rainwater collection, and
create a multi-level public green space and
green roof system. We can also learn from
policies of Zhongguancun Science Park
or set up a Zhongguancun Park branch
to attract high-tech enterprises to settle
in and to achieve win-win results through
school-enterprise cooperation and sharing
technology.

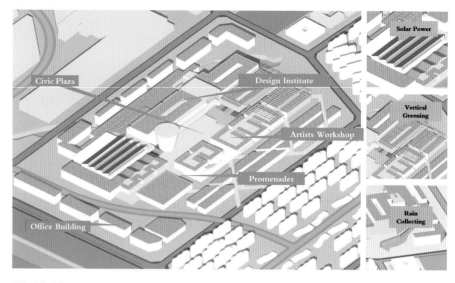

绿色建筑系统
Green building system

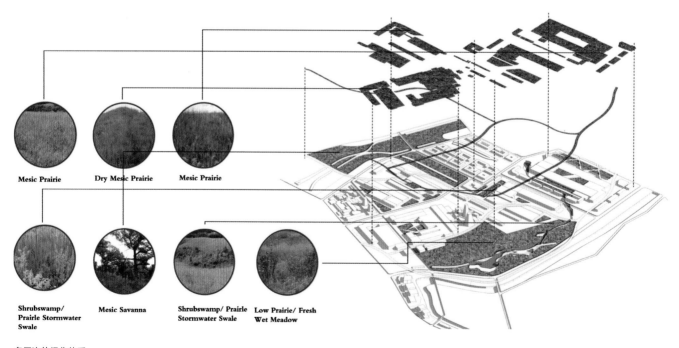

多层次的绿化体系
Multi-level green system

STRATEGY

CERAMIC WORKSHOPS ON BOTH SIDES OF GANGYAO ROAD

BUILDINGS INHERITING CULTURAL MEMORY

RETAIN TRADITIONAL CULTURE AND MEMORY

TRANSFORM IT INTO CULTURAL CENTER AND CULTURAL EXPERIENCING WORKSHOP

X3——文化创意
X3—Culture & creation

X3——文化创意
X3—Culture & creation

X3是绿+文化创意，地段的缸窑路周边有许多小型的陶瓷作坊，应保留这些建筑遗产、传承非物质文化遗产，将它们改造成文化中心和手工业技艺体验工作坊。工作坊临近大面积的绿色空间，将成为当地居民周末休闲娱乐的好去处。

X3 is Green+culture & creation. There are many small ceramic workshops on both sides of the Gangyao Road. We transform them into a culture center and a handcraft experience workshop. Adjacent to a large green space, they can be a great place of leisure for people to enjoy on weekends.

鼓励公众发挥创造力，"自下而上"地参与促进建筑遗产、非物质文化遗产的保留与传承。

Through encouraging public participation and the bottom-up preservation and inheritance of architectural heritage and intangible cultural heritage, we give full play to the creativity of people.

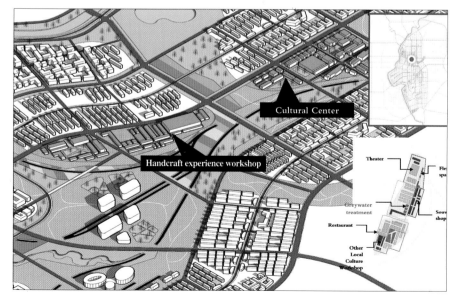

Cultural Center

Handcraft experience workshop

Theater

Greywater treatment

Restaurant

Souv shop

Other Local Culture Workshop

Fle spa

文化中心和手工业技艺体验工作坊
Culture center and handcraft experience workshop

X4——居住养老
X4—Residence & senior-care

X4是居住和养老，根据房屋质量判断居住社区的保留与拆除，规划面向休闲康养的住区，服务科研创新企业的住区和普通住区，并根据唐山人口规模制定分步建设计划。

X4 is Green+residence & senior-care. The residential communities on the site are preserved or dismantled according to their quality. We plan residential zones specially for seniors and the youths respectively, and ordinary ones according to needs of different groups of people. A step by step construction plan is made according to the size of population.

X4——居住养老
X4—Residence & senior-care

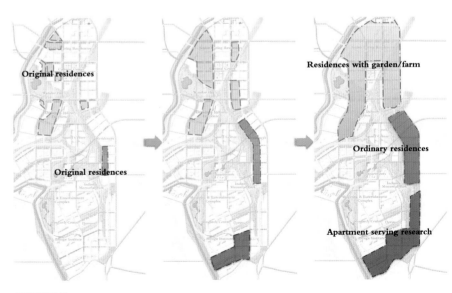

分布建设计划
Step by step construction

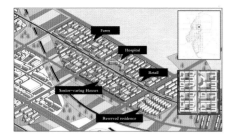

康养休闲住区
Residence for senior-care

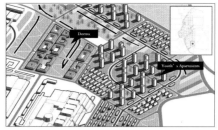

科技研发住区
Residence for researchers

康养产业发展政策
Policies for senior-care

康养休闲住区以合院式为主，园内布置农园或花园，作为休闲活动的承载空间。分布在科研创新区域周边的多是高层公寓，服务于这里数量众多的科技研发和教育产业从业人员。公共服务设施位于自行车绿廊两侧，包括餐厅、商店、学校和社区医院，为居民提供更便捷、绿色的生活。线性住区使生活在这里的居民最大化地接触到两侧的绿地和公共服务设施，做到"200米见绿"。政策方面，通过人才培养、顶层设计、多元融资实现对养老产业的全方位立体支持，通过税收减免、完善养老机构行业标准等措施，实现产业鼓励和严格监管。

These are courtyards that serve as senior-care. The center of the courtyard is a farm or garden, which is also a public place for leisure activities. The high-rise apartments and dormitories on the south side are targeted at the youth working in the surrounding research centers and schools. The bicycle lane in the middle of the settlement is a greenway for daily commuting. Public service facilities, which are available on both sides of the bicycle lane, include restaurants, shops, schools, and community hospitals, providing residents with a convenient life. This linear residential area maximizes access to the green and public services, so there can be green within every 200 meters. In terms of the policy, all-round support will be offered for the pension industry through training of professionals, policy design, and diversified financing; and industrial incentives and strict supervision are realized through measures, such as tax reduction and improvement of industry standards for pension institutions.

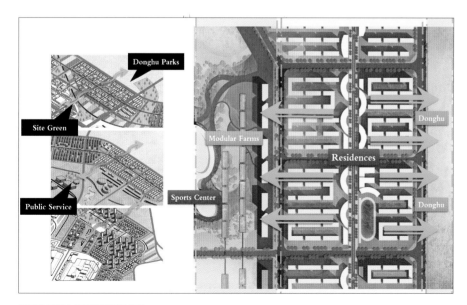

线形住区最大化接触两侧的绿色
Linear residence maximizes access to the green and public services on both sides

新的陡河东区将成为绿色基质中的宜居之岛，促成建设更绿色、更宜居的唐山。作为路北与开平的联络带，陡河东地段将改善二区目前生活工作品质低的问题，为唐山东北部带来新的活力。

The new Douhe East area in Tangshan will be a green substrate-based livable islet, making the city greener and more livable. This site, which is a place connecting Lubei and Kaiping districts, brings new vitality to the northeastern part of Tangshan.

地段南部将以极高的活力度形成新华路商务带终点，以渗透着景观的工业遗产塑造唐山工业文化特色形象，成为既符合年轻人审美潮流又能唤醒老居民工业记忆的城市新地标。陡河东将依托使用频繁、活力旺盛、触手可及的绿色基质，成为城市绿化改造先行区，为唐山市其他后工业区的升级改造奠定基础。

With high vitality, the southern part of the site will become the end of the Xinhua Road business belt. Through the regeneration of the industrial heritage sites and creating landscape, the image of Tangshan's industrial culture will be reshaped, and these heritage sites may become a new landmark of the city. The ecological transformation of the Douhe East area is a gift from Tangshan to the Beijing-Tianjin-Hebei Region, laying a foundation for the upgrading of other post-industrial areas in the city.

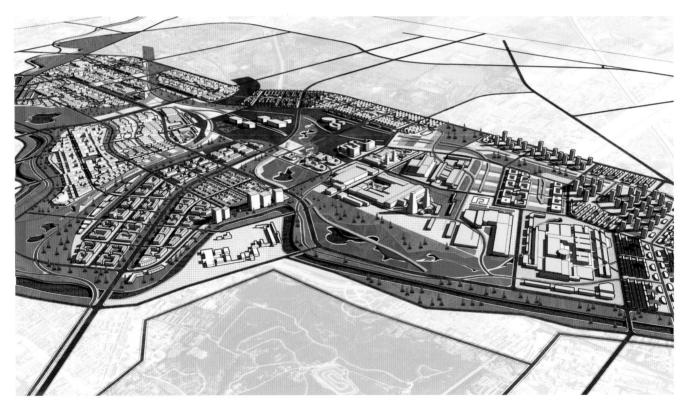

地段鸟瞰
Bird's eye view

心得体会
Summary

邵旭涛
SHAO Xutao

余旺仔
YU Wangzai

感谢课程中丰富的体验。第一次做如此大尺度的城市设计，我学到了与从前很不一样的规划和设计方法。在本课程设计学习与美国调研和交流中，我对后工业城市有了初步的印象和理解。因其历史遗留的布局、建筑可利用程度等问题，其改造升级仍为较大难题，也是城市更新中非常重要的类型。

Thanks to this studio, I had the first opportunity to do such a large-scale urban design, from which I gained a rich experience and learned a very different planning and design method. I also obtained a preliminary impression and understanding of post-industrial cities during my study in the US. Because of their historical problems, these cities still face challenges in urban regeneration.

和MIT同学的深度交流与合作让人难忘，在一学期的面对面接触和线上沟通中，我们彼此分享对地段的理解、新奇的概念，他们涌动的创新欲望让我赞叹和受益良多。感谢各位老师的悉心引导和慷慨包容，感谢队友和所有同学的朝夕相伴。

The in-depth exchanges and cooperations with MIT students are an unforgettable experience. Through the offline and online communications during the semester, we shared our understanding of the site and novel concepts. I benefit a lot from their marvelous innovative ideas. I am very grateful to our professors for their guidance and generosity and to classmates for their companionship.

附录
Appendix

2019清华大学—麻省理工学院城市设计联合工作坊大事记
A Chronicle of 2019 Tsinghua—MIT Urban Design Studio

2019年1月18日

◎ 2019年1月18日，MIT师生抵达北京，联合工作坊开始第一次在地联合工作。

18th January, 2019 MIT team arrived in Beijing. The first joint work of the studio started at Tsinghua University (THU).

2019年1月20日

◎ 2019年1月20日，联合工作坊正式启动。在简短的开幕式上，清华大学建筑学院党委书记张悦教授、清华—MIT城市设计联合工作坊资深负责人张杰教授、清华大学城市规划系副系主任王英副教授先后致辞，回顾了清华—MIT城市设计联合工作坊的合作历程，强调了国际合作教学对于推动城市学科发展的重要作用，并预祝本次联合工作坊取得成功；清华大学建筑学院唐燕副教授介绍了2019联合工作坊的教学安排和教学内容。在随后的学术讲座环节，麻省理工学院Brent Ryan教授以"收缩城市"为题，介绍了城市转型研究的国际学术前沿；清华大学建筑学院刘健副教授以"北京与京津冀协同发展"为题，介绍了中国的城市化进程以及区域视野中的京津冀协同发展趋势与规划；原唐山市规划局局长、清华大学中国城市研究院秘书长林澎以"唐山——中国城市转型的缩影"为题，介绍了唐山的城市发展历程和面临的现实问题；北京清华同衡规划设计研究院副院长胡洁教授结合其主持规划设计的两个重点项目，介绍了唐山南湖生态公园和唐山东湖森林公园的规划设计实践。

20th January, 2019 The studio was launched officially. At the brief opening ceremony, Professor Zhang Yue, Party Secretary of the School of Architecture of THU, Professor Zhang Jie, Senior Director of the Tsinghua—MIT Urban Design Studio, and Associate Professor Wang Ying, Deputy Head of the Department of Urban Planning of THU, delivered speeches successively. They reviewed the cooperation history of the Tsinghua—MIT Urban Design Studio, emphasized the significance of internationally cooperative teaching in promoting the development of urban planning discipline, and wished the 2019 Tsinghua—MIT Studio a great success. Associate Professor Tang Yan from the School of Architecture of THU introduced the arrangement and syllabus of the 2019 Tsinghua—MIT Studio. In the following academic lectures, Professor Brent Ryan of MIT introduced the international academic frontier of urban transformation study with the topic of *Shrinking City*. Associate Professor Liu Jian from the School of Architecture of THU introduced the urbanization process of China, as well as the coordinated development trend and planning progress of Beijing-Tianjin-Hebei Region from a regional perspective, with the topic of *Beijing and Coordinated Development of Beijing-Tianjin-Hebei*. Lin Peng, former Director of Tangshan Planning Bureau and Secretary General of the China Urban Research Institute of THU, introduced the urban development process and practical problems of Tangshan with the topic of *Tangshan-the Epitome of Urban Transformation in China*. Professor Hu Jie, Vice President of Tsinghua Tongheng Planning and Design Institute, introduced two planning and design practices under

his charge, which were Tangshan Nanhu Ecological Park and Tangshan Donghu Forest Park.

◎ 2019年1月21日，联合工作坊全体师生集体考察作为2022北京冬奥会和冬残奥会场馆所在地之一的首钢园区更新改造进展，随后前往唐山进行实地调研。

21st January, 2019 The two teams collectively investigated the urban regeneration project of Shougang Park in Beijing, one of the venues for the 2022 Beijing Winter Olympics and Paralympic Winter Games, and then set out for Tangshan for field study.

2019年1月22日

◎ 2019年1月22日，联合工作坊全体师生集体参观唐山市规划展览馆和唐山大地震纪念馆，听取唐山市自然资源和规划局副局长赵铁政对唐山市历次城市规划的介绍，并在当地专业人员陪同下，前往陡河以东的设计地段，针对唐山钢铁厂、陶瓷博物馆、陶瓷厂等重要城市节点进行现场踏勘。

22nd January, 2019 The two teams collectively visited the Tangshan Planning Exhibition Hall and the Tangshan Earthquake Memorial Hall. Zhao Tiezheng, Deputy Director of the Tangshan Natural Resources and Planning Bureau, introduced the urban planning history of Tangshan to the studio members. After that, accompanied by local professionals, all tutors and students headed to the design site, east of Douhe River, to conduct an on-the-spot investigation on important urban nodes like Tangshan Iron and Steel Plant, Ceramics Museum and Ceramic Factory.

2019年1月23日

◎ 2019年1月23日，联合工作坊全体师生分组，继续在陡河以东的设计地段进行场地踏勘。

23rd January, 2019 The students of the two teams were divided into four mixed groups, and conducted site studies respectively in the design area of east Douhe River.

2019年1月24日

◎ 2019年1月24日，联合工作坊全体师生集体调研唐山市工业博物馆、培仁里社区、汉斯故居、凤凰新城等重要城市更新项目，随后参加"转型中的唐山：国际经验与地方实践"学术研讨会，并在会后返京。研讨会由清华大学建筑学院唐燕副教授主持，唐山市自然资源和规划局局长高怀军致辞；麻省理工学院Brent Ryan副教授以"收缩转型城市的可持续发展"为题，介绍了美国在应对收缩城市领域进行的"自上而下"与"自下而上"相结合的设计实践；清华大学刘健副教授以"经济重构和城市更新的国际经验"为题，从产业革命的大趋势切入主题，介绍了德国鲁尔地区和巴黎东部地区的空间转型经验；上海交通大学王林教授以"上海工业区转型更新研究"为题，引介了上海的最新实践进展；中国科学院大学张路峰教授以"唐山陡河沿线城市更新的规划探讨"为题，探究了唐山市城市水资源与景观资源利用的潜力；唐山昱邦房地产开发有限公司林盛董事长以"棚户区改造现实与展望"为题，分析了唐山本地城市更新实践中的难点；天津大学城市规划设计研究院王旭春总工程师以"唐山文化空间格局"为题，探讨了唐山市工业转型中的工业遗产利用问题；华北理工大学刘嘉娜副教授以"唐山工业遗产现状调查及评价体系研究"

为题，介绍了对唐山工业遗产的摸底研究成果；清华大学建筑设计研究院四所李匡副所长以"北京老工业区更新实践"为题，分享了清华大学建筑设计研究院全过程参与北京首钢地区更新改造的体会；随后，与会专家以圆桌对话的形式展开讨论并与参会人员进行了问答交流。

24th January, 2019 The two teams collectively investigated the important urban projects in Tangshan, such as Tangshan Industrial Museum, Peirenli Community, Hans' Former Residence and Phoenix New City, and then participated in the academic seminar of *Tangshan in Transition: International Experience and Local Practice*, after which they returned to Beijing in the late afternoon. The seminar was hosted by Associate Professor Tang Yan from THU. Gao Huaijun, Director of Tangshan Natural Resources and Planning Bureau, made an address first of all. Associate Professor Brent Ryan of MIT introduced the top-down and bottom-up design practice of the United States in dealing with shrinking cities with the topic of *Sustainable Development of Shrinking Cities*. Professor Liu Jian of THU introduced the spatial transformation experience of Ruhr and eastern Paris against the background of industrial revolution in her speech of *International Experience of Economic Reconstruction and Urban Regeneration*. Professor Wang Lin from Shanghai Jiaotong University introduced the latest practical progress in Shanghai with the topic of *Transformation and Regeneration of Shanghai Industrial Areas*. Professor Zhang Lufeng from the University of Chinese Academy of Sciences explored the utilization potential of urban water and landscape resources in Tangshan with the topic of *Discussion on Urban*

Regeneration Planning Along the Douhe River in Tangshan. LinSheng, Chairman of Tangshan Yubang Real Estate Development Co., Ltd. analyzed the difficulties and institutional shackles in Tangshan's local urban regeneration practice with the topic of *Reality and Prospect of shantytown Renovation*. Wang Xuchun, the chief engineer of Tianjin University Urban Planning and Design Institute, discussed the utilization of industrial heritages during Tangshan's industrial transformation with the theme of *Cultural Spatial Pattern of Tangshan*. Associate Professor Liu Jiana of North China University of Science and Technology introduced her findings on Tangshan's industrial heritages with the topic of *Investigation and Evaluation System of Tangshan's Industrial Heritages*. Li Kuang, Deputy Director of Architectural Design and Research Institute of THU, shared his experience in participating in the whole regeneration process of Shougang with the topic of *Regeneration Practice of Beijing Old Industrial Areas*. In the end, all the experts joined the round-table discussion and answer questions from listeners.

2019年1月25日—28日

◎ 2019年1月25日—28日，联合工作坊的19名学生混合编组，分别从采矿与景观、基础设施、工业与遗产、单位住区与城中村等专题入手，针对设计地段进行资料汇总和分析，并在双方教师指导下，完成初步的设计研究成果。

25th-28th January, 2019 The students of the two teams were divided into mixed groups and started to collect and analyze the data of the design area from various perspectives, such as mining and landscape, infrastructure, industry and heritage, Danwei communities and urban villages, following which the preliminary research outcomes were completed under the guidance of tutors from both sides.

2019年1月29日

◎ 2019年1月29日，联合工作坊举办第一次在地联合设计评图。各组同学分别汇报了各自的专题研究和方案设计初步成果，随后清华大学朱文一教授和毛其智教授、中国科学院大学张路峰教授、麻省理工学院Lorena Bello Gomez副教授进行了点评。

29th January, 2019 The first studio review was held. Each mixed group reported their preliminary outcomes of thematic research and design schemes, followed by comments from Professor Zhu Wenyi and Professor Mao Qizhi of THU, Professor Zhang Lufeng from Chinese Academy of Sciences, and Associate Professor Lorena Bello Gomez from MIT.

2019年1月30日

◎ 2019年1月30日，MIT师生离京返美，联合工作坊开始异地线上联合工作。

30th January, 2019 MIT team left Beijing for the United States. The studio entered the phase of individual design workshop, while the two teams kept regular online communications.

2019年4月12日

◎ 2019年4月12日，联合工作坊全体师生开展异地线上联合设计评图。

12th April, 2019 The second studio review was held online.

2019年4月27日

◎ 2019年4月27日，清华团队离京赴美，联合工作坊开始第二次在地联合工作。

27th April, 2019 Tsinghua team left Beijing for the United States, and the studio entered the phase of joint design workshop for the second time.

2019年4月28日

◎ 2019年4月28日，联合工作坊全体师生实地调研波士顿城市建设，考察贝肯山、码头市场、Big Dig、SOWA等重点城市更新项目。

28th April, 2019 The two teams collectively conducted a field inspection on Boston's urban construction, and visited several key urban regeneration projects in Boston, such as Beacon Hill, Wharf Market, Big Dig and SOWA.

2019年4月29日

◎ 2019年4月29日，清华师生与MIT专家学者开展学术交流，MIT房地产研究中心访问学者Kairos Shen、Design X项目执行主任Gilad Rosenzweig、感知城市实验室Snoweria Zhang和Viktorija Abolina分别介绍了关于城市再开发、设计交叉研究、城市数据应用和参与式城市更新的前沿研究成果。

29th April, 2019 Tsinghua team had academic exchanges with MIT experts and scholars. Visiting scholar Kairoshen of MIT Real Estate Research Center, Gilad Rosenzweig, Executive Director of Design X Project, Snoweria Zhang and Viktorija Abolina of MIT Senseable City Lab, introduced respectively the cutting-edge studies on urban redevelopment, cross-disciplinary design study, urban data application and participatory urban regeneration.

2019年4月30日

◎ 2019年4月30日，清华师生实地考察Lowell、Lawrence两座传统工业城市的转型发展，并与当地规划部门、社区组织、企业机构的代表进行座谈，共同探讨工业建筑改造利用等城市更新实践的多元途径。

30th April, 2019 Tsinghua team visited two traditional American industrial cities, Lowell and Lawrence, to study their transformation and development. They had discussions with representatives of local planning departments, community organizations and enterprises to explore various approaches for urban regeneration practice, such as the transformation and utilization of industrial buildings.

2019年5月1日

◎ 2019年5月1日，联合工作坊的19名学生分组完善各自的汇报成果。

1st May, 2019 The students of the two teams worked in mixed groups to improve their design and presentation work.

2019年5月2日

◎ 2019年5月2日，联合工作坊第二次在地联合设计评图，17位来自美国学界和业界的顶级专家对学生的设计研究成果进行了点评和建议，包括MIT荣退教授、宾夕法尼亚大学设计学院前院长Gary Hack教授，MIT房地产研究中心主任、Design X项目负责人、城市设计与开发项目原主任Dennis Frenchman教授，MIT规划系主任Eran Ben Joseph教授，MIT后城市主义研究中心联合主任、阿卡汗讲席教授Jim Wescoat和城市设计与规划讲师Marie Law Adams，MIT建筑系Sheila Kennedy教授、Andrew Scott教授、Rafi Segal副教授和讲师Yao Zhang，MIT市政数据设计实验室主任Sarah Williams副教授，MIT建筑系访问讲师Susanne Schindler，哈佛大学Raymond Garbe讲席教授Peter Row和Martin Bucksbaum讲席教授Joan Busquets，哈佛大学荣退资深研究员Peter Del Tredici，康涅狄格大学景观建筑学项目主任Jill Desimini副教授，Sasaki规划与设计公司合伙人Dennis Piertz，Utile建筑与规划事务所创始人Tim Love。

2nd May, 2019 The studio launched the second joint design evaluation. 17 top experts from American academic and industry circles gave comments and suggests to the design works of students, including Professor Gary Hack, former Dean of the School of Design of the University of Pennsylvania and Professor Emeritus of MIT; Professor Dennis Frenchman, Director of MIT Real Estate Research Center, project leader of Design X, and former director of urban design and development project; Professor Eran Ben Joseph, Director of the Department of Urban Studies and Planning of MIT; Jim Wescoat, Co-director of the MIT Post-urbanism Research Center, Aga Khan Professor Emeritus; Dr. Marie Law Adams, principal of Somerville, Massachusetts, firm Landing Studio; Professor Sheila Kennedy, Professor Andrew Scott, Associate Professor Rafi Segal, and Lecturer Yao Zhang, from the Department of Architecture of MIT; Associate Professor Sarah Williams, Director of MIT Municipal Data Design Laboratory;

Visiting Lecturer Susanne Schindler, Raymond Garbe Professor Peter Row, and Martin Bucksbaum Professor Joan Busquets, from Harvard University; Peter Del Tredici, Senior Fellow Emeritus of Harvard University; Associate Professor Jill Desimini, Director of Landscape Architecture Program of University of Connecticut; Dennis Piertz, Partner of Sasaki Planning and Design Company, and Tim Love, Founder of Utile Architecture and Planning Office.

2019年5月3日—8日

◎ 2019年5月3日—8日，清华师生实地考察美国部分城市的城市更新项目。

3rd-8th May, 2019 Tsinghua team visited urban regeneration projects in some other cities in the United States.

2019年5月9日

◎ 2019年5月9日，清华师生离美返京，联合工作坊开始异地独立工作阶段。

9th May, 2019 Tsinghua team returned to Beijing from the United States, and the two teams began to work independently.

2019年6月2日

◎ 2019年6月2日，清华师生独立进行联合工作坊最终设计评图，邀请唐山市自然资源

152

和规划局高怀军局长、唐山市昱邦慈善公益基金会创始人林盛、北京清华同衡规划设计研究院副总规划师胡洁、清华大学中国城市研究院秘书长林澎、清华大学建筑学院党委书记张悦教授、前任院长朱文一教授、前任书记边兰春教授、张杰教授、教学办主任钟舸副教授作为评图嘉宾，对同学们的设计研究成果进行点评。

12th June, 2019 Tsinghua tutors and students conducted the independent final design evaluation. 9 Experts have been invited to comment the students'design and research results, including Gao Huaijun, Director of Tangshan Natural Resources and Planning Bureau, Lin Sheng, Founder of Tangshan Yubang Charity Foundation, Hu Jie, Deputy Chief Planner of Beijing Tsinghua Tongheng Planning and Design Institute, Lin Peng,

former Director of Tangshan Planning Bureau and Secretary General of the China Urban Research Institute of THU, Professor Zhang Yue, Party Secretary of the School of Architecture of THU, and Professor Zhu Wenyi, Professor Bian Lanchun, Professor Zhang Jie and Associate Professor Zhong Ge.

2019年6月3日—14日

◎ 2019年6月3日—14日，联合工作坊最终成果在清华大学建筑学院展出。

3rd-14th June, 2019 The final results of the joint studio were exhibited at the School of Architecture of Tsinghua University.